Green Energy and Technology

Climate change, environmental impact and the limited natural resources urge scientific research and novel technical solutions. The monograph series Green Energy and Technology serves as a publishing platform for scientific and technological approaches to "green"—i.e. environmentally friendly and sustainable—technologies. While a focus lies on energy and power supply, it also covers "green" solutions in industrial engineering and engineering design. Green Energy and Technology addresses researchers, advanced students, technical consultants as well as decision makers in industries and politics. Hence, the level of presentation spans from instructional to highly technical.

Indexed in Scopus.

Indexed in Ei Compendex.

Xiaolin Wang
Editor

Future Directions in Energy Engineering

Challenges, Opportunities, and Sustainability

Springer

Editor
Xiaolin Wang
School of Engineering
University of Tasmania
Hobart, TAS, Australia

ISSN 1865-3529 ISSN 1865-3537 (electronic)
Green Energy and Technology
ISBN 978-3-031-62041-6 ISBN 978-3-031-62042-3 (eBook)
https://doi.org/10.1007/978-3-031-62042-3

This Springer imprint is published by the registered company Springer Nature Switzerland AG
The registered company address is: Gewerbestrasse 11, 6330 Cham, Switzerland

Contents

Exploring Weather Research and Forecasting (WRF) Model Parameterizations for Enhanced Solar and Wind Energy Resources Predictions in Malawi

Sylvester William Chisale, Han Soo Lee, and Manuel Soto Calvo

1 Introduction

The increasing emphasis on renewable energy (RE) sources is a global priority, and the accuracy of weather predictions plays a pivotal role in this transition. Unfortunately, developing countries such as Malawi have limited weather stations to mimic solar and wind energy potentials in a particular region. Furthermore, financial challenges can hinder installation of weather stations across the country [1]. Thus, with poor electrification rate in Malawi [2] and global increasing use of RE, precise weather forecasting is essential for optimizing the performance of solar and wind energy systems. The Weather Research and Forecasting (WRF) Model [3], an open-source atmospheric simulation system, is instrumental in predicting weather conditions.

The increasing use of downscaling tools, such as the WRF model, could greatly benefit developing countries, especially in assessing solar and wind energy

S. W. Chisale (✉)
Transdisciplinary Science and Engineering Program, Graduate School of Advanced Science and Engineering, Hiroshima University, Hiroshima, Japan

Department of Applied Studies, Malawi University of Science and Technology (MUST), Limbe, Malawi

H. S. Lee
Transdisciplinary Science and Engineering Program, Graduate School of Advanced Science and Engineering, Hiroshima University, Hiroshima, Japan

Center for Planetary Health and Innovation Science (PHIS), The IDEC Institute, Hiroshima University, Hiroshima, Japan

M. S. Calvo
Transdisciplinary Science and Engineering Program, Graduate School of Advanced Science and Engineering, Hiroshima University, Hiroshima, Japan

X. Wang (ed.), *Future Directions in Energy Engineering*, Green Energy and Technology, https://doi.org/10.1007/978-3-031-62042-3_1

resources. Among these tools, WRF-solar has been notably enhanced for comprehensive solar energy application [4]. These models have been applied in various regions for both offshore and onshore applications, as well as solar resource assessments such as Arabian Peninsula [4] and Cambodia [5].

In Malawi, weather stations, mainly used for agriculture, may not fully capture wind energy potential. The WRF model addresses challenges like non-standard wind speed measurements at 2 m and data gaps from poor maintenance. Also, the region lacks sensitivity studies for solar and wind speed predictions using the WRF model.

Solar and wind energy, increasingly affordable and scalable, are vital for future national energy planning. Thus, the study mainly examined the sensitivity of the WRF-solar model to shortwave (SW) irradiance parameterization for assessing solar energy resources under varying seasonal and sky conditions, including aerosol impacts. It also investigated the WRF model's sensitivity to wind speed for wind energy assessment, considering different parameterization schemes like microphysics, land surface physics, planetary boundary layer (PBL), and cumulus schemes.

2 Methodology

2.1 Study Location

Malawi, located in the southern part of Africa covers about 118,484 km^2. It is characterized by varied topography including highlands and lake Malawi. The country experiences a subtropical climate with varied weather patterns due to its diverse topography and altitude differences, as depicted in Fig. 1, which also shows the distribution of weather stations (data from the Department of Climate Change and Meteorological Services).

2.2 Model Setup

The study employed WRF model [3] version 4.4.1 for wind parameterization and WRF-solar version 4.5.1 for solar radiation parameterization, both developed by institutes associated with the National Center for Atmospheric Research (NCAR) [6]. The models used two nested domains: d01 (12 km grid spacing, 104 × 133 grid points, 38 vertical levels) and d02 (4 km grid spacing, 121 × 268 grid points, 38 vertical levels). The domain setup used in this study is shown in Fig. 1. The study utilized ERA5 reanalysis data [7] and further explore NCEP final operational global surface analyses data. Both models utilized cumulus scheme Kain-Fritsch [8] for outermost domain and no cumulus scheme for innermost (<10 km resolution) [9].

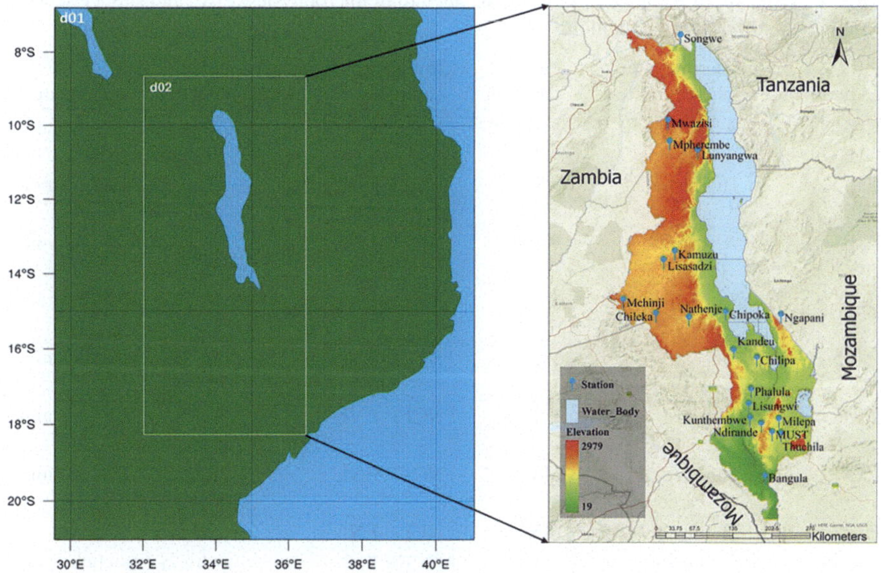

Fig. 1 Domain setup for Malawi including elevation profile and the distribution of stations

2.3 Sensitivity Experiments

Several numerical experiments were conducted to investigate the sensitivity of wind speed at 10 m high and GHI predictions. For wind speed parameterization, 13 sensitivity studies were conducted over an eight-day period from 15 to 23 January 2022. The experiments also compared the performances of NCEP FNL and ERA5 reanalysis data as initial and boundary conditions. Microphysics, land surface physics, PBL, and cumulus schemes were considered for wind parameterization. Table 1 illustrates the wind speed experimental setup conducted over Malawi.

For GHI parameterization, 12 sensitivity studies were conducted over an eight-day period with priority to Thompson aerosol microphysics to establish the aerosol effects on solar radiation. Thus, the experiments were focused on optimal SW radiation scheme influencing GHI predictions. Five schemes were evaluated: Dudhia, the updated Goddard scheme, the Community Atmosphere Model (CAM), the updated Rapid Radiative Transfer Model (RRTM), and the Goddard Fluid Dynamics Laboratory (GFDL). Two different seasons were considered, the rainy (15–23 January 2022) and sunny seasons (06–14 August 2022). Furthermore, the experiment for GHI prediction considered utilizing Goddard microphysics scheme which has not aerosol effect. Table 2 illustrates the experimental setup conducted over Malawi.

Table 1 Wind speed parameterization setups, including microphysics, PBL, and surface schemes with initial and boundary conditions

Experiments	Microphysics schemes	PBL schemes	Initial & boundary condition	Surface layer schemes
Exp1	WSM3	YSU	NCEP-FNL	Revised MM5 Monin-Obukhov
Exp2	Thompson	YSU	NCEP-FNL	Revised MM5 Monin-Obukhov
Exp3	WSM6	YSU	NCEP-FNL	Revised MM5 Monin-Obukhov
Exp4	Morrison	YSU	NCEP-FNL	Revised MM5 Monin-Obukhov
Exp5	Lin	YSU	NCEP-FNL	Revised MM5 Monin-Obukhov
Exp6	Goddard	YSU	NCEP-FNL	Revised MM5 Monin-Obukhov
Exp7	Thompson	MYNN2	NCEP-FNL	Revised MM5 Monin-Obukhov
Exp8	Thompson	MYNN3	NCEP-FNL	MYNN
Exp9	Thompson	MYJ	NCEP-FNL	Monin-Obukhov (Janjic Eta)
Exp10	Goddard	MYNN2	NCEP-FNL	Revised MM5 Monin-Obukhov
Exp11	Goddard	MYJ	NCEP-FNL	Monin-Obukhov (Janjic Eta)
Exp12	Goddard	MYNN3	NCEP-FNL	MYNN
Exp13	Goddard	YSU	ERA5	Revised MM5 Monin-Obukhov

2.4 Evaluation Procedure

Model performances for both wind and GHI were compared with observational datasets using correlation coefficient (CC), mean absolute error (MAE), root mean squared error (RMSE), and standard deviation error (STDE), in line with previous studies [5, 10]. Thus, Eqs. (1)–(4) were used to determine CC, MAE, RMSE, and STDE for all n variables in both the model (M) and the observation (O) whereas $\overline{O_i}$ and $\overline{M_i}$ are their respective means. For STDE, regardless of a high RMSE or Bias, a low STDE indicates consistent error, indicating acceptable simulation physics [11]. The wind speed measurements were taken at a 2 m height, with some stations reporting missing data. For model evaluation, wind speed was extrapolated to 10 m using the wind power law, $v = v_r(h/h_r)^{\alpha}$, where α is power law index, v_r is the reference wind speed at reference height, h_r. In this study, $\alpha = 0.47$ was adopted based on a nearby study [12].

Table 2 GHI setups: Exp 1–5 (blue) focus on shortwave radiation experiments under constant conditions

Experiment	Parameterization	Scheme
	Microphysics	Thompson aerosol
	Radiation (Longwave)	RRTMG
	Surface Layer	Revised MM5 (Revived Monin-Obukhov)
	Land-Surface	Noah
	Planetary Boundary Layer (PBL)	Mellor-Yamada-Janjic (MYJ)
	Cumulus	Kain-Fritsch (12 km only)
Exp1	Radiation (Shortwave)	Dudhia
Exp2		New Goddard
Exp3		CAM
Exp4		RRTMG
Exp5		GFDL
Exp6	Microphysics	Goddard
	Radiation (Longwave)	RRTM
	Surface Layer	Revised MM5 (Revived Monin-Obukhov)
	Land-Surface	Noah
	Planetary Boundary Layer (PBL)	Mellor-Yamada-Janjic (MYJ)
	Cumulus	Kain-Fritsch (12 km only)
	Radiation (Shortwave)	Dudhia

Exp6 (red) involves Goddard microphysics, no aerosol (optimized for wind)

$$\mathrm{CC} = \frac{\sum_{i=1}^{n}\left(M_i - \overline{M_i}\right)\left(O_i - \overline{O_i}\right)}{\sqrt{\sum_{i=1}^{n}\left(M_i - \overline{M_i}\right)^2 \sum_{i=1}^{n}\left(O_i - \overline{O_i}\right)^2}} \tag{1}$$

$$\mathrm{MAE} = \frac{1}{n}\sum\nolimits_{i=1}^{n}\left|M_i - O_i\right| \tag{2}$$

$$\mathrm{RMSE} = \left[\frac{1}{n}\sum\nolimits_{i=1}^{n}\left(M_i - O_i\right)^2\right]^{\frac{1}{2}} \tag{3}$$

$$\mathrm{STDE} = \left[(\mathrm{RMSE})^2 - (\mathrm{Bias})^2\right]^{\frac{1}{2}} \tag{4}$$

3 Results and Discussions

3.1 Optimal Model Setup Determination

The study conducted a total of 23 experiments for both GHI and wind speed. For wind speed, 12 weather station data were evaluated out of 13 experiments, and we computed average statistical parameters to determine the final configuration. Figure 2a illustrates outperforming model combination using the Taylor diagram. Thus, Exp13 is the optimal combination for wind speed prediction using WRF model over Malawi. The optimal model combination includes Goddard microphysics, YSU PBL scheme, MM5 similarity scheme for surface layer, no cumulus for the inner domain, Kain-Fritsch for the larger domain, and ERA5 reanalysis data for initial and boundary conditions.

Experiment with GHI for both cloudy and sunny days, from the shortwave radiation sensitivity including changing microphysics scheme, Exp5 has proven to be optimal model combination in both seasons. Thus, the optimal combination includes Thompson aerosol microphysics scheme, the updated Rapid Radiative Transfer (RRTM) scheme for longwave, the Goddard Fluid Dynamics Laboratory (GFDL) scheme for shortwave, the Revised Monin-Obukhov (Revised MM5) surface layer scheme and the unified Noah land-surface scheme. Furthermore, for PBL is the Mellor-Yamada-Janjic (MYJ) scheme whereas for cumulus, the Kain-Fritsch scheme for larger domain and no cumulus for innermost domain.

The model performance also showed some improvements on sunny days as compared to cloudy days as depicted in Fig. 2a, b. Furthermore, by considering microphysics scheme which does not consider aerosol effects which happened to be the optimal setup for wind parameterization output, the WRF-solar model did not perform well as compared to considering aerosol effect using Thompson aerosol microphysics scheme.

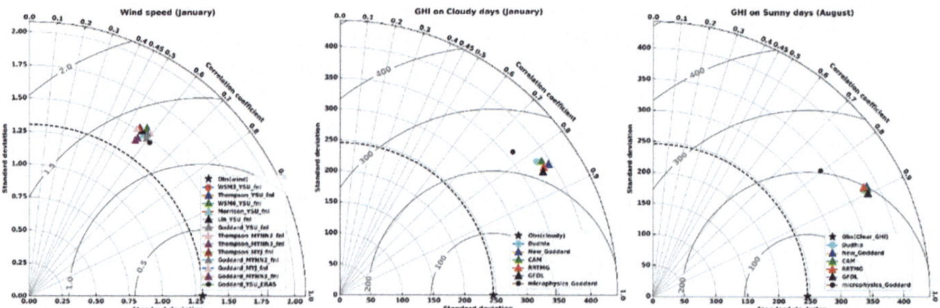

Fig. 2 Taylor diagram illustrating the optimal model parameterization with (**a**) wind speed, (**b**) GHI on cloudy days and (**c**) GHI on sunny days

3.2 *Optimal Model Evaluation*

This paper evaluated experiments for wind speed using statistical metrics (CC, RMSE, MAE, and STDE). Different sites exhibited varying performances in each experiment; however, despite these variations, the overall performance at other stations was consistently good across all experiments. The statistical performances presented in Fig. 3 (optimal setup for wind) showed that most metrics were similar, with the best performances at Lisungwi EPA, which demonstrated a strong correlation of 0.86. In contrast, Mwazisi and Kandeu performed poorly. Similarly, when comparing GHI statistics for both cloudy and sunny days, Fig. 3 (optimal setup for solar) indicated that the CC for GHI was very good compared to that for wind speed. Furthermore, sunny days generally showed more improved CC values than cloudy days. However, in both wind and GHI comparisons, the study revealed large error metrics. These errors in wind speed parameterization could be attributed to interpolation errors in the wind speed data (2 m anemometer height) and poor maintenance of weather stations, as evidenced by significant amounts of missing data in both datasets, including long periods of missing data.

3.3 *Long Term Simulation Results*

Based on the optimal model configurations, the optimal configurations were simulated for 1 year period. For wind energy assessment, wind speed was plotted at 10 m and 50 m (using WRF-Python for interpolation [13]) as depicted in Fig. 4. The spatial analysis reveals higher wind speeds at higher elevations, beneficial for large wind turbines. While average wind speeds are generally low, some areas exhibit speeds suitable for energy generation, reaching up to 10 m/s. Considering the entire domain, average wind speeds are 3.77 m/s at 10 m and 4.1 m/s at 50 m height. For

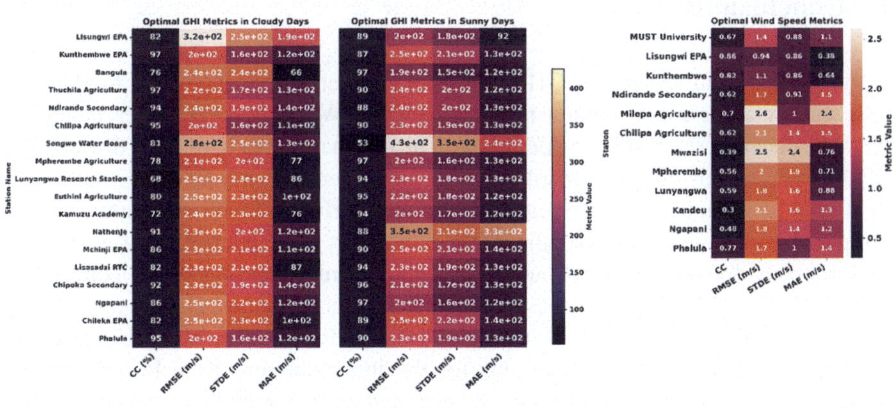

Fig. 3 Wind speed and seasonal GHI statistical performances at different stations

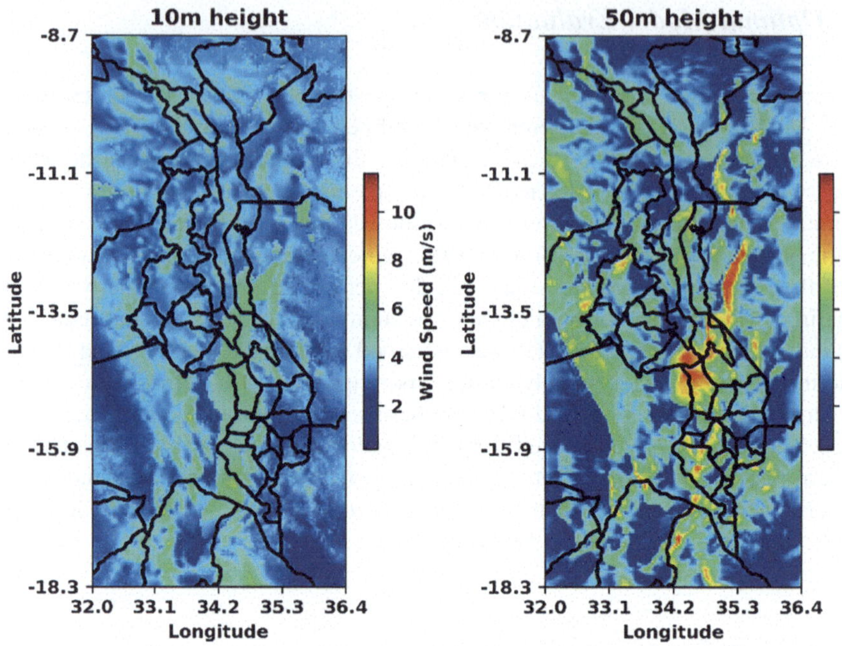

Fig. 4 Mean wind speed spatial plot at 10 and 50 m height for the year 2022

solar energy assessment, considering temperature and GHI, the spatial plot (Fig. 5) illustrates high GHI values in most parts of the southern region with also high values over the north part of the lake. Thus, the GHI values can reach as high as 8 kWh/m^2/day. Similarly, high temperature values are depicted over the lake and the southern region of Malawi.

4 Conclusions

In this study, wind speed and GHI data were evaluated utilizing WRF model. For solar energy assessment, emphasis was given to SW radiation schemes including aerosol effect and seasonal considerations. The optimal SW radiation for the site was GFDL. The inclusion of aerosols further improved the model performance as compared to without it. Additionally, the model performed well on sunny days as compared to cloudy days. On the other hand, for wind energy assessment, the analysis considered NCEP FNL and ERA5 data as initial and boundary conditions, microphysics, land surface physics, PBL, and cumulus schemes. Hence, the optimal model setup consists of Goddard microphysics, YSU PBL scheme, MM5 similarity scheme for surface layer, no cumulus for the inner domain, Kain-Fritsch for the larger domain, and ERA5 reanalysis data for initial and boundary conditions. The one-year mean wind speed at 50 m revealed some potential sites for wind energy

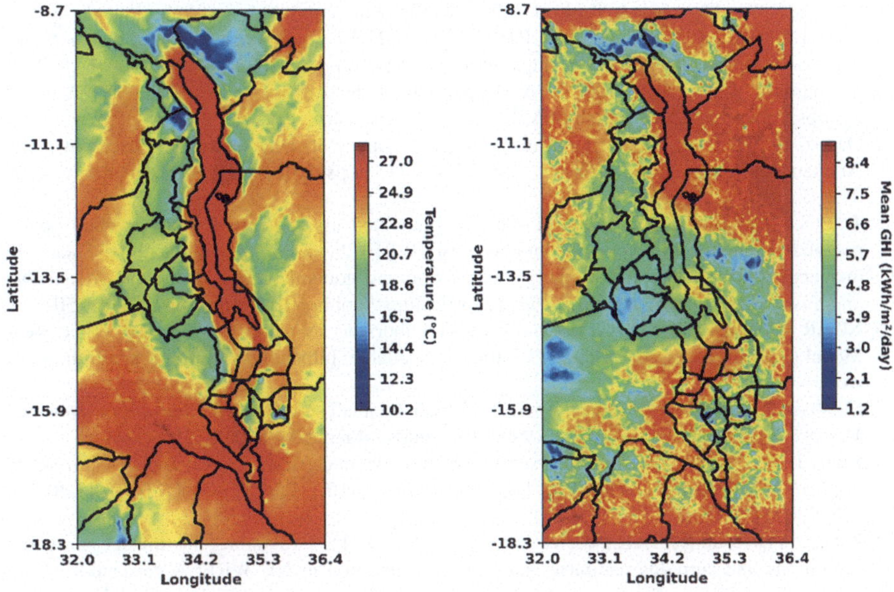

Fig. 5 Mean temperature and GHI for the simulated year 2022

generation with entire domain's annual mean of 4.1 m/s. Furthermore, the one-year simulation revealed significant potential solar energy in Malawi with annual GHI values as high as 8 kWh/m^2/day. Larger values were seen in most parts of across the northern part of the lake and over much of the southern region.

In summary, the application of the WRF model in Malawi significantly enhances the prediction of solar and wind energy resources, fostering collaborative RE development. Its results are crucial in strategic planning for RE, including grid integration and microgrid projects. Furthermore, given the data limitations in developing countries like Malawi, it becomes a valuable tool for these nations in pursuing sustainable energy solutions.

Acknowledgements We are grateful to the Japan International Cooperation Agency (JICA) for funding the first and third authors through the SDGs Global Leadership Program, the Malawi Department of Meteorological Services for crucial data, and the CHESS Lab for insightful contributions.

References

1. Chisale, S. W., & Lee, H. S. (2023). Evaluation of barriers and solutions to renew-able energy acceleration in Malawi, Africa, using AHP and fuzzy TOPSIS approach. *Energy for Sustainable Development, 76*, 101272. https://doi.org/10.1016/j.esd.2023.101272

2. Chisale, S. W., Eliya, S., & Taulo, J. (2023). Optimization and design of hybrid power system using HOMER pro and integrated CRITIC-PROMETHEE II approaches. *Green Technologies and Sustainability, 1*(1), 100005. https://doi.org/10.1016/j.grets.2022.100005
3. Skamarock, W. C., et al. (2008). A Description of the Advanced Research WRF Version 3. https://doi.org/10.5065/D68S4MVH
4. Dasari, H. P., et al. (2019). High-resolution assessment of solar energy resources over the Arabian peninsula. *Applied Energy, 248*, 354–371. https://doi.org/10.1016/j.apenergy.2019.04.105
5. Tuy, S., Lee, H. S., & Chreng, K. (2022). Integrated assessment of offshore wind power potential using weather research and forecast (WRF) downscaling with Sentinel-1 satellite imagery, optimal sites, annual energy production and equivalent CO_2 reduction. *Renewable and Sustainable Energy Reviews, 163*, 112501. https://doi.org/10.1016/j.rser.2022.112501
6. NCAR. Mesoscale & Microscale Meteorology Laboratory. Weather Re-search & Forecasting Model (WRF). Accessed 5 Oct 2023 [online]. Available: https://www.mmm.ucar.edu/models/wrf
7. Hersbach, H., et al. (2020, July). The ERA5 global reanalysis. *Quarterly Journal of the Royal Meteorological Society, 146*(730), 1999–2049. https://doi.org/10.1002/qj.3803
8. Kain, J. S. (2004). The Kain–Fritsch convective parameterization: An update. *Journal of Applied Meteorology, 43*(1), 170–181. https://doi.org/10.1175/1520-0450(2004)043<0170:TKCPAU>2.0.CO;2
9. Zhang, Y., Roundy, J. K., & Santanello, J. A. (2021). Evaluating the impact of model resolutions and cumulus parameterization on precipitation in NU-WRF: A case study in the central Great Plains. *Environmental Modelling & Software, 145*, 105184. https://doi.org/10.1016/j.envsoft.2021.105184
10. Dayal, K. K., Bellon, G., Cater, J. E., Kingan, M. J., & Sharma, R. N. (2021). High-resolution mesoscale wind-resource assessment of Fiji using the weather research and forecasting (WRF) model. *Energy, 232*, 121047. https://doi.org/10.1016/j.energy.2021.121047
11. Carvalho, D., Rocha, A., Gómez-Gesteira, M., & Santos, C. (2012). A sensitivity study of the WRF model in wind simulation for an area of high wind energy. *Environmental Modelling & Software, 33*, 23–34. https://doi.org/10.1016/j.envsoft.2012.01.019
12. Mwanyika, H. H., & Kainkwa, R. M. (2006). Determination of the power law exponent for southern highlands of Tanzania. *Tanzania Journal of Science, 32*(1), 103–108. https://doi.org/10.4314/tjs.v32i1.18434
13. NCAR. wrf-python. Accessed 5 Oct 2023 [online]. Available: https://wrf-python.readthedocs.io/en/latest/

Intraday Solar Irradiance Forecasting Based on Hybrid Machine Learning Methodology for Photovoltaic Power Applications

Ibrahim M. Gaber and Rania A. Ibrahim

1 Introduction

Solar energy is gaining popularity as a renewable energy source due to its ease of installation, maintenance, and operation. However, maintaining balance between consumption and production is crucial for power production plants. Accurate supply prediction is essential for PV systems, as it offers benefits for power grid planning, operation, and management [1].

Solar irradiance is a crucial parameter for determining solar unit characteristics and estimating potential power output in regions, leading to extensive research in recent years [2]. Solar irradiation forecasters may be characterized into two major dimensions: the prediction horizon; and the model utilized to build the forecaster. Regarding time horizon, four types of forecasters exist: (a) very short-term forecasters, that provide predictions for the next few minutes play a vital role in real-time dispatch [3]; (b) short-term forecasters, which generate predictions ranging from several hours up to a day ahead and are mainly involved in unit-commitment or economic dispatch [4]; (c) medium-term forecasters produce predictions ranging from a day to few weeks ahead, and these prediction can be used in maintenance planning [5]; and (d) long-term forecasters, whose predictions timeframe can range from a few weeks to months ahead and may be employed in power assessment and new infrastructure development.

As for the second dimension which pertains the models adopted as solar irradiation forecasters, forecasters may be classified as physical, statistical, or artificial intelligence-based models [6]. Physical models use numerical weather prediction, statistical models use historical databases, and artificial intelligence models build

I. M. Gaber (✉) · R. A. Ibrahim
Electrical and Control Engineering Department, Arab Academy for Science, Technology and Maritime Transport, Alexandria, Egypt
e-mail: I.Mohamed16165@student.aast.edu; rania_assem@aast.edu

X. Wang (ed.), *Future Directions in Energy Engineering*, Green Energy and Technology, https://doi.org/10.1007/978-3-031-62042-3_2

nonlinear correlations between input and output parameters [7–9]. Research on solar irradiance forecasting focuses on statistical and physics-based models for short-term and long-term predictions, impacting power scheduling decisions and PV system integration strategies [10].

In this regard, this paper explores the forecasting abilities of different ML regression-based models, including SVMR, ANN, and DT for solar irradiance prediction. Additionally, the paper further investigates the possible fusion of these models for an elevated prediction accuracy and enhanced performance. The constructed models are evaluated and tested for short term solar irradiance forecast using historic meteorological dataset. Through implementation and comparison, this work seeks to discover the most suited algorithms for solar irradiance forecasting taking into account historical weather conditions.

2 Solar Irradiance Predictive Models

Intra-hour forecasting methods are divided into univariate and multivariate models [11]. Univariate models rely on historical data for predicting future irradiance levels, but struggle with high fluctuations. Multivariate models consider multiple input variables, like meteorological variables, for solar irradiance prediction. For this reason, meteorological variables are commonly used, such as temperature, humidity, wind speed, etc. for solar irradiance prediction which can be corelated to PV power production [12].

As mentioned earlier, artificial intelligence based on ML models have been widely used to improve the estimation of forecast. According to existing literature, several studies have been undertaken for forecasting solar irradiance, typically using individual ML models, or using ensemble of different models. In Aslam et al. (2019) [13], authors explored how several ML models might be used in predicting daily and hourly solar radiation. The results showed an improved performance when these models were fed with meteorological data as recommended by Khosravi et al. (2018) [14]. The authors [15] developed two forecasting ML models based on two types of the input data; one which uses local time, temperature, pressure, wind speed, and relative humidity as input variables of the models. The second model uses time-series prediction of solar irradiance using multilayer feed-forward neural network (MLFFNN). The work of Yagli et al. [16] investigates the hourly forecasting performance of 68 ML algorithms for three sky conditions, seven locations, and five climate zones for development of a universal model. More recently, ensemble methods have increasingly been used to improve algorithms' prediction performance since they combine multiple algorithms into one to make an enhanced predictor. For forecasting long-term solar radiation, Aslam et al. (2020) [17] combined auto-encoder (AE) and Long Short-Term Memory (LSTM) models. Kumari and Toshniwal (2021) [18] suggested a stacking ensemble model fusing XGBT with Deep Neural Networks (DNNs). The work of AlKandari and Ahmad (2019) [19] presented a hybrid approach that blended statistical methodology with ML models for forecasting solar generation using simple averaging and weighted averaging.

In highlight of the above, results of previous studies showed that ensemble models offer more promising prediction performance compared to individual regressor models, which set the main objective of this work.

2.1 Individual ML-Based Models

ML regression approaches can be widely divided into ANN-based models, kernel-based SVMR techniques, tree-based models, and ensembled hybrid models.

ANN-Based Models ANN-based models, inspired by biological neural systems, utilize neurons in three layers for processing information, outperforming traditional empirical models in linear and nonlinear problems [20]. The ANN output can be expressed as (1):

$$y(x) = F\left(\sum_{i=1}^{j} \omega_i(t).x_i(t) + b\right) \quad (1)$$

where $x_i(t)$ are the input variables, $y(x)$ is the output, F is the transfer function of the hidden layer, j is the number of neurons; $\omega_i(t)$ and b are the weight and the bias, respectively.

SVMR Models Support vector machines (SVM) are kernel-based machine learning models used for classification and regression problems, utilizing kernel functions to map low-dimensional data into high-dimensional feature spaces. The objective function of SVM can be expressed as [21]:

$$f(x) = \omega\varphi(x) + b \quad (2)$$

Such that $\varphi(x)$ is the higher dimensional feature space. Though SVMR presented excellent performances in prediction, it is reported that their computational cost is high.

Tree-Based Models DT, or non-parametric hierarchical structures, are efficient for handling large variables and cases. Regression trees divide training data into smaller subsets, reducing training time and computational burden. DT's general equation can be represented as follows (3):

$$\sum\nolimits_{j=1}^{j} \sum_{i\epsilon R_j} \left(\text{pred}_{R_j} - y_i\right)^2 \quad (3)$$

where, pred_{R_j} is observation mean sum of the jth branch.

2.2 Hybrid-Based Models

Hybrid and ensembled model generation aims to improve accuracy and robustness by combining hypotheses from multiple base learners, unlike individual models that learn a single hypothesis.

In this work, simple arithmetic mean of each individual model prediction is carried out. Combining ML can be achieved via simple averaging or applying weighted average. In the simple averaging approach, the final forecasted irradiation values are calculated by averaging the forecast generated from individual ML models, as in (4) [22]:

$$\widehat{y} = \sum_{k=1}^{m} \widehat{y_k}/m \tag{4}$$

where $\widehat{y}$ is the final predicted value of irradiance, m defines the number of ML algorithms in the ensemble, $\widehat{y}_k$ represents the predicted values of the k algorithm.

3 Forecasting Methodology

The proposed methodology for hour ahead solar irradiance forecasting uses six ML-based approaches: SVMR, DT, ANN, and their ensembled combination. Data is pre-processed with cleaning, normalization, and filter-based feature selection modules for standardization as shown in Fig. 1. Using the min-max transformation approach, each feature is linearly rescaled to the [0, 1] interval by the min-max normalizer as presented by (5):

$$z = \frac{n - \min(n)}{[\max(n) - \min(n)]} \tag{5}$$

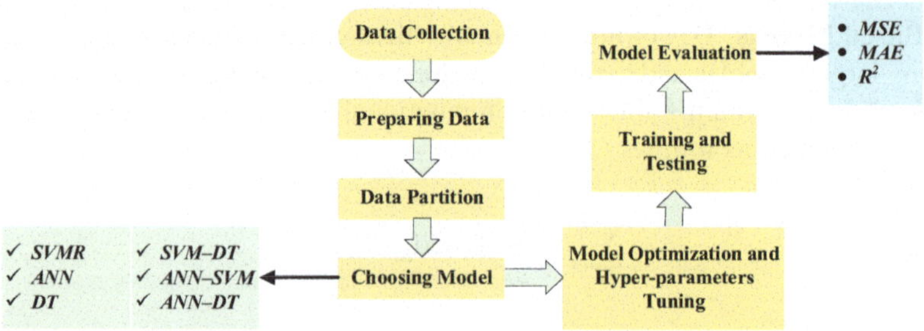

Fig. 1 Forecasting methodology of solar radiation using machine learning algorithms

where n is the original number, and min and max value of n in order to compute min-max transformation.

A filter-based feature selection is the last stage of data preprocessing. This stage is crucial for building the ML algorithms as it identifies the input dataset columns with the best solar radiation prediction abilities.

Next stage is data partitioning and splitting into training, validation, and test sets. The data set used in this work is a public dataset comprising 79,913 observations totaling minute-by-minute readings of the ambient temperature, wind speed, and solar irradiance collected from actual weather station data from a solar plant for the complete year of 2020 [23]. The study used 70% and 30% data divisions for training and testing ML algorithms, with 70% training and 30% testing, comparing performance after training and testing. The remaining untrained data was then employed to evaluate the model's performance after utilizing the trained dataset.

As mentioned previously, individual models using SVMR, DT, and ANN are first tested, followed by the ensembled hybrid models after hyper-parameter tuning. Model optimization and hyper tuning of parameters stage is essential in ML so that its model performance is maximized especially with large datasets and high-dimensional parameter spaces [20]. Following this stage, models are trained and tested, subsequently models' evaluation using performance metrics are carried out.

For the validation and evaluation phase, the accuracy of the ML models studied in this work were evaluated using widely known performance metrics, mean square error (MSE), mean average error (MAE), and coefficient of determination (R^2) as (6–8). These statistical indices serve as critical benchmarks in quantifying how effectively ML models forecast, ensuring the reliability and precision of the calculated predictions.

$$\text{MAE} = \frac{1}{h}\sum\nolimits_{k=1}^{h}(y_k - \widehat{y}_k) \tag{6}$$

$$\text{MSE} = \frac{1}{h}\sum\nolimits_{k=1}^{h}\left(Y_k - \widehat{Y}_k\right)^2 \tag{7}$$

$$R^2 = \frac{\sum_{k=1}^{h}(y_k - \overline{Y}_k)^2 - \sum_{i=1}^{h}(y_k - \widehat{y}_k)^2}{\sum_{i=1}^{h}(y_k - \overline{y}_k)^2} \tag{8}$$

where h represents the number of observations. Low MAE and MSE values, along with R^2 values approaching unity, indicate accurate and reliable estimations.

4 Results & Performance Evaluation

The study evaluates the forecasting accuracy of individual machine learning algorithms using a dataset, then constructs hybrid models using averaging approach, aiming to find the best model with highest prediction accuracy. Figures 2 and 3 show

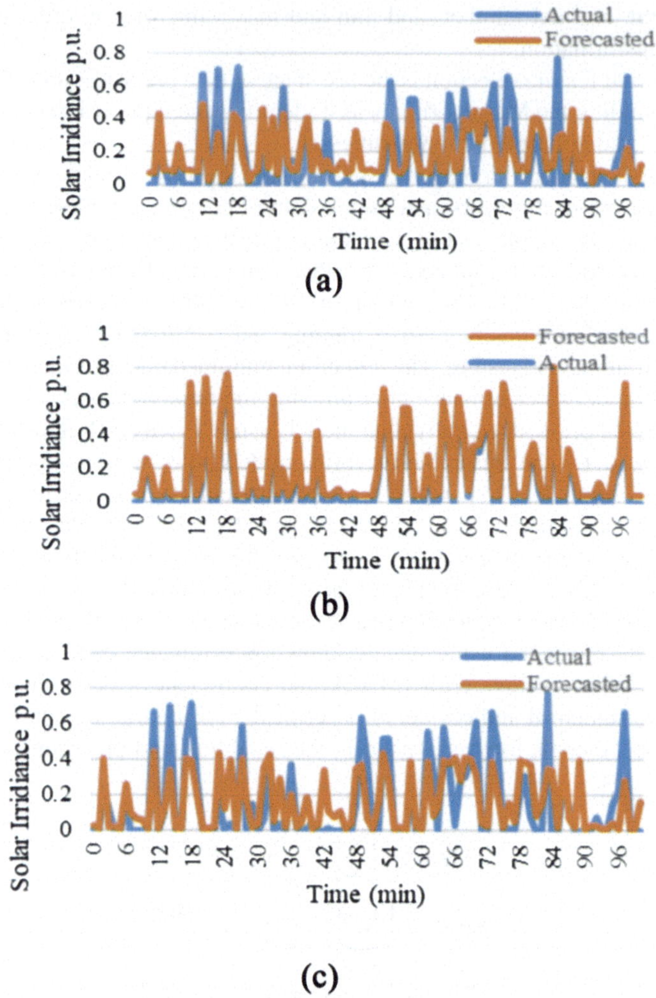

Fig. 2 Actual and predicted solar irradiance values (**a**) SVMR, (**b**) DT, and (**c**) ANN

a correlation between actual and predicted results from three ML models, but determining the best model is challenging. Performance parameters (6–8) are used for qualitative comparison.

Table 1 displays the decomposition of the results obtained for each of the tested individual ML models for solar irradiation forecasting using MAE, MSE, and R^2. It can be seen from Table 1 that the MSEs of the SVMR, DT, and ANN individual algorithms are, respectively, 0.0285, 0.027, and 0.026. These figures pertain to the accuracy of the algorithm. It has been observed that ANN achieves the highest level of prediction accuracy, followed by DT and SVMR. Similarly, by comparing the MAE results, DT achieves the lowest score of 0.089, followed by ANN, and the

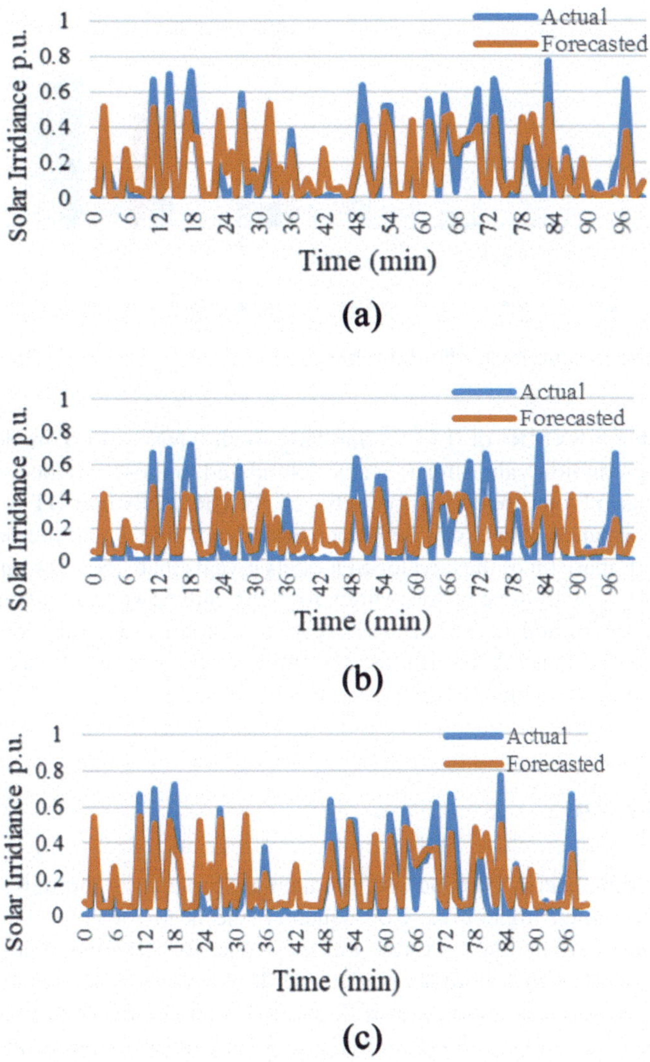

Fig. 3 Actual and predicted solar irradiance values (**a**) ANN–DT, (**b**) ANN–SVMR, and (**c**) SVMR–DT

Table 1 Performance metrics of different models

	SVMR	ANN	DT	ANN-SVM	ANN-DT	SVMR-DT
MSE	0.0285	0.026	0.027	0.0271	0.0215	0.0223
MAE	0.1301	0.114	0.089	0.1234	0.0976	0.1069
R^2	0.4467	0.491	0.491	0.4752	0.5835	0.5676

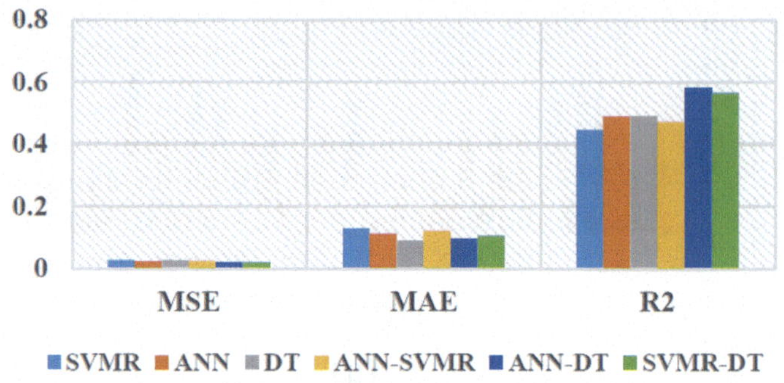

Fig. 4 Summary of error results for ML models

highest MAE for SVMRs of 0.1301. Similarly, both ANN and DT models produce better fitted prediction values for R^2 values compared to those produced by SVMR.

For the ensembled models, fusion between SVMR, ANN and DT is carried out and the performance results can be demonstrated in Table 1 and the bar charts of Fig. 4. The comparison of the developed models reveals that ANN-DT outperformed other ensembles, with lower error metrics for MSE and MAE, and highest R^2 score of 0.0215, 0.0976, and 0.5835, respectively. It is noteworthy that ANN-DT also excelled the individual ML algorithms. The summary of error results can be depicted in Fig. 4 for a more comprehensive graphical presentation.

5 Conclusion

With the ongoing energy demand rise, producing clean energy through solar photovoltaics is becoming imperative. Consequently, the need for precise and accurate solar irradiance prediction becomes indispensable for effective integration and utilization of PV energy within the power grid. In this work, a comparative intraday forecasting performance analysis of different ML models has been presented. For this purpose, base learners (SVRM, ANN, and DT) and their potential ensembled models are compared and tested using temporal database from actual weather data. In terms of achieving the highest estimated accuracies, the obtained results showed that ensembled strategies in general outperformed the individual base learner models.

References

1. Haider, S. A., Sajid, M., Sajid, H., Uddin, E., & Ayaz, Y. (2022, October). Deep learning and statistical methods for short- and long-term solar irradiance forecasting for Islamabad. *Renewable Energy, 198*, 51–60. https://doi.org/10.1016/j.renene.2022.07.136

2. Jalali, S. M. J., Ahmadian, S., Kavousi-Fard, A., Khosravi, A., & Nahavandi, S. (2022, January). Automated deep CNN-LSTM architecture Design for Solar Irradiance Forecasting. *IEEE Transactions on Systems, Man, and Cybernetics: Systems, 52*(1), 54–65. https://doi.org/10.1109/TSMC.2021.3093519
3. Bouzgou, H., & Gueymard, C. A. (2019, April). Fast short-term global solar irradiance forecasting with wrapper mutual information. *Renewable Energy, 133*, 1055–1065. https://doi.org/10.1016/j.renene.2018.10.096
4. David, M., Luis, M. A., & Lauret, P. (2018, July). Comparison of intraday probabilistic forecasting of solar irradiance using only endogenous data. *International Journal of Forecasting, 34*(3), 529–547. https://doi.org/10.1016/j.ijforecast.2018.02.003
5. Alonso, J., & Batlles, F. J. (2014, August). Short and medium-term cloudiness forecasting using remote sensing techniques and sky camera imagery. *Energy, 73*, 890–897. https://doi.org/10.1016/j.energy.2014.06.101
6. Rodríguez, F., Martín, F., Fontán, L., & Galarza, A. (2021, August). Ensemble of machine learning and spatiotemporal parameters to forecast very short-term solar irradiation to compute photovoltaic generators' output power. *Energy, 229*, 120647. https://doi.org/10.1016/j.energy.2021.120647
7. Dong, L., Wang, L., Khahro, S. F., Gao, S., & Liao, X. (2016, July). Wind power day-ahead prediction with cluster analysis of NWP. *Renewable and Sustainable Energy Reviews, 60*, 1206–1212. https://doi.org/10.1016/j.rser.2016.01.106
8. Caldas, M., & Alonso-Suárez, R. (2019, December). Very short-term solar irradiance forecast using all-sky imaging and real-time irradiance measurements. *Renewable Energy, 143*, 1643–1658. https://doi.org/10.1016/j.renene.2019.05.069
9. Kushwaha, V., & Pindoriya, N. M. (2019, September). A SARIMA-RVFL hybrid model assisted by wavelet decomposition for very short-term solar PV power generation forecast. *Renewable Energy, 140*, 124–139. https://doi.org/10.1016/j.renene.2019.03.020
10. Wojtkiewicz, J., Hosseini, M., Gottumukkala, R., & Chambers, T. L. (2019, October). Hour-ahead solar irradiance forecasting using multivariate gated recurrent units. *Energies (Basel), 12*(21), 4055. https://doi.org/10.3390/en12214055
11. Rana, M., Koprinska, I., & Agelidis, V. G. (2016, August). Univariate and multivariate methods for very short-term solar photovoltaic power forecasting. *Energy Conversion and Management, 121*, 380–390. https://doi.org/10.1016/j.enconman.2016.05.025
12. Shan, S., Li, C., Ding, Z., Wang, Y., Zhang, K., & Wei, H. (2022, October). Ensemble learning based multi-modal intra-hour irradiance forecasting. *Energy Conversion and Management, 270*, 116206. https://doi.org/10.1016/j.enconman.2022.116206
13. Aslam, M., Lee, J. M., Kim, H. S., Lee, S. J., & Hong, S. (2019). Deep learning models for long-term solar radiation forecasting considering microgrid installation: A comparative study. *Energies (Basel), 13*(1), 147. https://doi.org/10.3390/en13010147
14. Khosravi, A., Koury, R. N. N., Machado, L., & Pabon, J. J. G. (2018). Prediction of hourly solar radiation in Abu Musa Island using machine learning algorithms. *Journal of Cleaner Production, 176*, 63. https://doi.org/10.1016/j.jclepro.2017.12.065
15. Khosravi, A., Koury, R. N. N., Machado, L., & Pabon, J. J. G. (2018, March). Prediction of hourly solar radiation in Abu Musa Island using machine learning algorithms. *Journal of Cleaner Production, 176*, 63–75. https://doi.org/10.1016/j.jclepro.2017.12.065
16. Yagli, G. M., Yang, D., & Srinivasan, D. (2019, May). Automatic hourly solar forecasting using machine learning models. *Renewable and Sustainable Energy Reviews, 105*, 487–498. https://doi.org/10.1016/j.rser.2019.02.006
17. Aslam, M., Lee, J. M., Altaha, M. R., Lee, S. J., & Hong, S. (2020). AE-LSTM based deep learning model for degradation rate influenced energy estimation of a PV system. *Energies (Basel), 13*(17), 4373. https://doi.org/10.3390/en13174373
18. Kumari, P., & Toshniwal, D. (2021). Extreme gradient boosting and deep neural network based ensemble learning approach to forecast hourly solar irradiance. *Journal of Cleaner Production, 279*, 123285. https://doi.org/10.1016/j.jclepro.2020.123285

19. AlKandari, M., & Ahmad, I. (2019). Solar power generation forecasting using ensemble approach based on deep learning and statistical methods. *Applied Computing and Informatics, 16*(2), 1–20. https://doi.org/10.1016/j.aci.2019.11.002
20. Gaber, I. M., & Ibrahim, R. A. (2023, June). Hourly electricity Price prediction applying deep learning for electricity market management. In *2023 IEEE international conference on environment and electrical engineering and 2023 IEEE industrial and commercial power systems Europe (EEEIC / I&CPS Europe)* (pp. 1–5). IEEE. https://doi.org/10.1109/EEEIC/ICPSEurope57605.2023.10194867
21. Zhou, Y., Liu, Y., Wang, D., Liu, X., & Wang, Y. (2021, May). A review on global solar radiation prediction with machine learning models in a comprehensive perspective. *Energy Conversion and Management, 235*, 113960. https://doi.org/10.1016/j.enconman.2021.113960
22. An, K., & Meng, J. (2010). *Voting-Averaged Combination Method for Regressor Ensemble* (pp. 540–546). https://doi.org/10.1007/978-3-642-14922-1_67
23. GitHub – pcsingh/Time_Series_Forecasting: Time series forecasting of solar irradiance data. https://github.com/pcsingh/Time_Series_Forecasting. Accessed 14 Sept 2023

Integrated Energy Management System Approach for Off-Grid Residential Home

Tolulope Olumuyiwa Falope, Liyun Lao, and Da Huo

1 Introduction

According to the United Nations, fossil fuels like coal, oil, and gas make up more than 75% of greenhouse gas (GHG) emissions worldwide and almost 90% of carbon dioxide emissions overall; fossil fuels are by far the biggest cause of climate change [1]. To mitigate against this, the share of renewable energy sources (RES) in the grid needs to be increased [2]. Non-dispatchable or variable renewable energy (VRE) sources such as wind and solar are intermittent and create stability and reliability issues as penetration levels increase [3]. Solar PV generation has seen an uptick in recent years. According to the International Energy Agency (IEA), solar PV generation experienced the largest growth of all RES in 2022 and reached almost 1300 TWh, surpassing wind for the first time [4]. To significantly increase solar energy integration, the existing grid requires efficient demand and grid management, adequate regulatory frameworks, updated operation and planning design, improved grid flexibility, energy storage, and supportive and enabling technologies to better match supply to demand [5]. One such planning and management procedure is the integrated energy management system (IEMS). IEMS just like integrated energy systems (IES) coordinates the planning and optimal operation of multi-energy systems using predictive and real-time controls to initiate both supply and demand responses in balancing the load and power supply in the grid.

T. O. Falope (✉) · L. Lao · D. Huo
Energy and Sustainability, School of Water, Energy & Environment, Cranfield University, Bedford, UK
e-mail: t.o.falope@cranfield.ac.uk

X. Wang (ed.), *Future Directions in Energy Engineering*, Green Energy and Technology, https://doi.org/10.1007/978-3-031-62042-3_3

In reviewing the state-of-the art, Huang et al. [6] develop a mathematical model to simulate flexible loads. The authors use genetic algorithm (GA) to solve the optimization problem of minimizing electricity purchase cost by matching renewable energy generation with electricity demand. Huy et al. [7] propose a supervised-learning model that simultaneously schedules the charging and discharging of energy storage systems (ESS) and electric vehicles (EVs) while also forecasting the energy demand. Kumar and Karthikeyan [8] propose an intelligent Golden Jackal Optimization (GJO) model to minimize operating costs by managing batteries and hybrid energy sources (HESs). This is achieved while considering constraints such as energy storage performance, power balance, generation capacity, and consumer loads. In all the proposed methods, there is no IEMS that uniquely combines SEF, TOU, GC, and DLC to optimize solar energy integration.

The key objective of this study is to investigate how the proposed IEMS prioritizes PV and reduces CO_2 emissions by optimally scheduling and dispatching PV and DG power supply. This is achieved with the help of a three-step SEF model, and the IEMS algorithm that initiates TOU, GC and DLC responses.

2 Methods

2.1 PV Sizing Optimization Using the Performance Ratio Method

Optimally sizing the PV system ensures sufficient supply capacity and reduces waste while fully utilizing the available PV generation. The Performance Ratio (PR) PV sizing method suggested by Isiolaotan et al. [9] uses a PR value of 65% and the peak sun hours (PSH) of the system location. The PR value of 65% is the system efficiency minus losses. The PSH is the amount of hours in a day in which the solar irradiance of the sun produces 1000 W per square meter [10]. The average PSH in the UK is between 2 and 3 h a day. The first step determines the total battery discharge by calculating the maximum or average daily energy used and the inverter standby load. The solar energy demand and the solar array power are then determined by the following equations:

$$\text{Solar Energy Demand} = \frac{\text{Total Battery Discharge}}{\text{PR}} \tag{1}$$

$$\text{Solar Array Power} = \frac{\text{Solar Energy Demand}}{\text{PSH}} \tag{2}$$

2.2 Three-Step Forecasting Approach

A three-step forecasting approach for forecasting PV generation is proposed. This approach identifies weather parameters with moderate to strong positive correlation to solar radiation, from both an on-site and local weather station, and combines them with low-data fusion to predict solar energy generation. This approach has been extensively analyzed in Falope et al. [11].

2.3 IEMS Framework

The IEMS receives as its input a day-ahead solar forecast of the PV system from the proposed three-step forecasting architecture approach described above. Based on the forecast, it sets the TOU tariffs/and or load schedule. For predictive control, the IEMS with the help of smart meter control, can set higher TOU tariffs to reflect periods of low PV supply, notify users about low and peak periods so that they can change/update load priority requests (manually schedule loads), or automatically schedule the load based on PV availability. By doing this, the IEMS determines the operating schedule of shiftable and interruptible household appliances for the next day-ahead.

For real-time control, the IEMS manages the system by checking if the current demand exceeds the supply level. If this condition is met, the IEMS's decisive algorithm will turn-off appliances via smart sockets/plugs by using DLC to reduce the demand based on a priority order determined by the user. If this is still not sufficient to curtail the load, then the back-up generator will be called upon to increase the supply. Figure 1 shows the flowchart of the proposed IEMS.

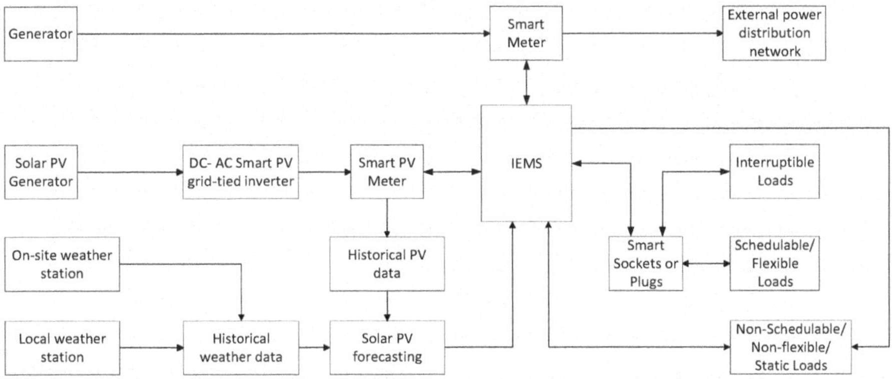

Fig. 1 Block diagram of proposed IEMS

2.4 Data Description

One-year historical Bedford weather data (from both an onsite and local weather station), one-year historical PV generation, and one-year hourly residential load consumption are obtained for the case study.

The daily load is categorized into base load, shiftable load, and interruptible load. Interruptible loads (plug loads) are loads that can be briefly switched off without causing the user too much discomfort [12]. Shiftable loads (heater and water pump) can be moved to an alternate time of operation [13], and base loads cannot be scheduled or interrupted. Table 1 gives the load summary along with the average operating times.

3 Results and Discussion

To size the system, we obtain the maximum daily energy demand from Table 1. The size of the inverter is determined by summing all the individual loads and multiplying it by a factor of 1.2. This approximately equates to 7.4 kVA, so a 7.5 kVA inverter with a 98% efficiency is selected. Applying Eqs. (1) and (2), the system size was determined to be 15.4 kW. A 15-kW roof-top array located in Cranfield University, with access to the historical PV generation data, along with the historical weather data are used to determine the next-day PV generation using the three-step forecast approach. With the next day hourly PV generation and the annual load consumption, the IEMS plans and schedules the load and supply. The results are displayed in Table 2.

Table 1 Load demand summary of residential house

Load type	Load	Ratings (W)	Qty	Average operating time (h/day)	Total demand (Wh/day)
Base	Lights	10	8	6	480
	TV	360	1	8	2880
	Fridge	240	1	6	1440
	Ceiling fan	80	3	9	2160
Shiftable	Heater	3000	1	2	6000
	Water pump	800	1	1	800
	Heater	3000	1	2	6000
Interruptible	Plug loads	200	Lump	9	1800
	Total	**4690**			**15,560**

Table 2 Consumption comparison between DG, DG/PV, and DG/PV/IEMS configurations for a residential house in 2019

	DG (kWh)	DG/PV (kWh)		DG/PV/IEMS (kWh)	
Month	DG load	DG load	PV load	DG load	PV load
Jan	734.61	701.41	33.20	542.44	109.32
Feb	590.08	517.60	72.48	356.31	186.01
Mar	487.94	393.25	94.69	258.49	205.33
Apr	446.91	369.90	77.01	213.99	222.00
May	485.81	344.41	141.40	173.16	306.86
Jun	470.14	311.44	158.70	172.50	292.50
Jul	485.81	314.05	171.76	167.43	314.66
Aug	485.81	352.07	133.74	175.33	301.14
Sep	455.85	377.32	78.53	213.15	239.97
Oct	485.81	423.20	62.61	277.82	182.21
Nov	455.85	422.54	33.31	323.33	88.90
Dec	485.81	459.05	26.76	360.51	70.96
Total	6070.44	4986.24	1084.20	3234.46	2519.86
Average	505.87	415.52	90.35	269.54	209.99

Table 2 shows that DG/PV has a higher combined load generation output (energy supplied from both DG and PV) than the DG/PV/EMS configuration. This is because the IEMS shuts down some interruptible loads without rescheduling them. In addition, the IEMS reduces the yearly generator output by about 47%, leading to reduced CO_2 emissions and fuel costs. The Scope 1 emissions, which are direct emissions, is calculated by multiplying the UK's average electricity CO_2 conversion factor (0.25 $kgCO_2e/kWh$) [14] by the total kWh used by the DG in the year, and the system lifespan which is 20 years. Using this formula, the total CO_2 for DG, DG/PV, and DG/PV/IEMS are 30,352.18 kg CO_2e, 24,931.19 kg CO_2e, and 16,172.32 kg CO_2e respectively. This means the IEMS reduces the CO_2 emissions compared to the DG and DG/PV configurations by 47% and 35% respectively. Even more importantly, the IEMS improves the yearly PV usage of the DG/PV configuration by more than 132%, thus achieving the overall aim of integrating more PV into the grid system. Figure 2 gives some further insight into the impact of the IEMS on the system.

In the first graph, the DG follows the load closely. Even though excess PV is generated between 7 am and 12 pm, the major load consumption occurs outside this time frame, between the hours of 4 am–7 am and 5 pm–9 pm. This means that the excess PV must be exported. In the second graph, the IEMS shifts the load peaks to better coincide with the maximum PV generation. As a result, there is an increase in load usage as can be seen by the bigger load peaks during 7 am and 12 pm. In addition, these shifted loads are increasingly being powered by PV.

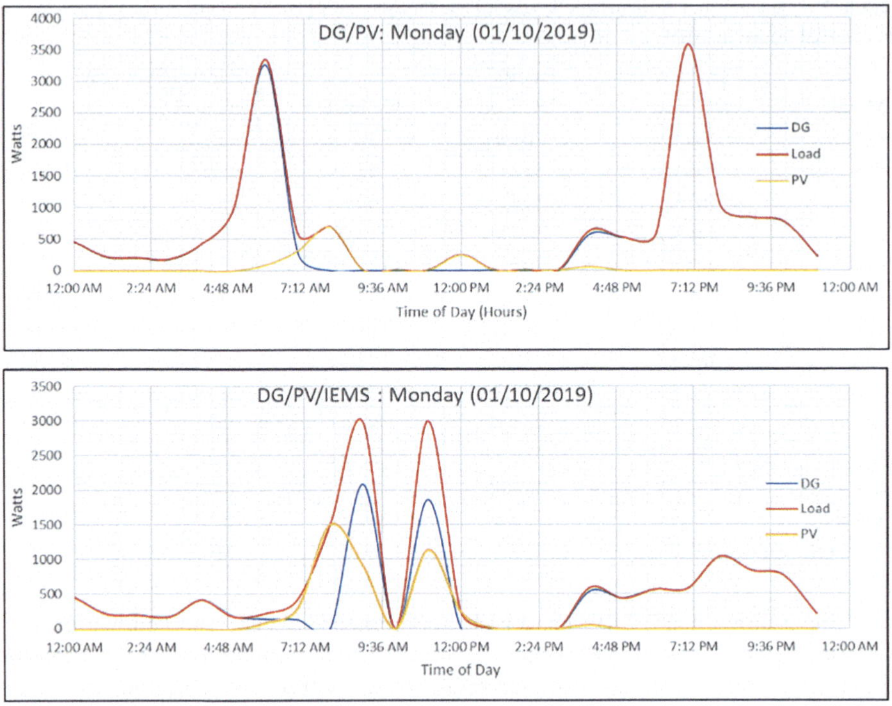

Fig. 2 Comparison of DG/PV and DG/PV/IEMS configurations

4 Conclusion

An Integrated Energy Management System (IEMS) that combines demand-supply side management was proposed with the objective of managing and increasing photovoltaic (PV) usage in the grid. This method integrates solar energy forecasting (SEF), time-of-use (TOU), direct load control (DLC), and generator control (GC), prioritizing load and energy management before calling up additional power supply. The performance of the IEMS was investigated by comparing three configurations: diesel generator (DG), DG/PV and DG/PV/IEMS. The DG/PV/IEMS showed an increase PV usage over the DG/PV by 132%. The IEMS also reduced the CO_2 emissions by 47% and 35% in the DG and DG/PV configurations respectively.

The impact of the IEMS can be further investigated by increasing the share of interruptible and schedulable loads and investigating other demand-side and supply-side management techniques.

Credit Author Statement **Tolulope O. Falope**: conceptualization, methodology, software, data curation, formal analysis, visualization, writing—original draft; **Lao Liyun**: writing—review & editing, supervision; **Da Huo**: writing—review & editing, supervision.

Declaration of Interests The authors declare that they have no known competing financial interests or personal relationships that could have appeared to influence the work reported in this paper.

Funding This research was supported by the Petroleum Technology Development Fund (PTDF) [PTDF/ED/OSS/PHD/TOF/1945/20].

References

1. United Nations. (2023). *Causes and effects of climate change*. United Nations. https://www.un.org/en/climatechange/science/causes-effects-climate-change. Accessed 7 Dec 2023.
2. Hainsch, K., Löffler, K., Burandt, T., Auer, H., Crespo del Granado, P., Pisciella, P., et al. (2022). Energy transition scenarios: What policies, societal attitudes, and technology developments will realize the EU Green Deal? *Energy, 239*, 122067. https://doi.org/10.1016/J.ENERGY.2021.122067
3. IEA. (2021). *World energy outlook 2021.*
4. IEA. (2023). *Solar PV.* International Energy Agency.
5. IRENA. (2015). *Renewable energy integration in power grids.*
6. Huang, Z., Wang, F., Lu, Y., Chen, X., & Wu, Q. (2023). Optimization model for home energy management system of rural dwellings. *Energy, 283*, 129039. https://doi.org/10.1016/J.ENERGY.2023.129039
7. Huy, T. H. B., Truong Dinh, H., Ngoc Vo, D., & Kim, D. (2023). Real-time energy scheduling for home energy management systems with an energy storage system and electric vehicle based on a supervised-learning-based strategy. *Energy Conversion and Management, 292*, 117340. https://doi.org/10.1016/J.ENCONMAN.2023.117340
8. Kumar, R. P., & Karthikeyan, G. (2024). A multi-objective optimization solution for distributed generation energy management in microgrids with hybrid energy sources and battery storage system. *Journal of Energy Storage, 75*, 109702. https://doi.org/10.1016/J.EST.2023.109702
9. Isiolaotan, O., Androsch, H., & Stadter, H. (2017). *Solar photovoltaic installation* (3rd ed.). Deutsche Gesellschaft für Internationale Zusammenarbeit (GIZ) GmbH.
10. Megantoro, P., Syahbani, M. A., Sukmawan, I. H., Perkasa, S. D., & Vigneshwaran, P. (2022). Effect of peak sun hour on energy productivity of solar photovoltaic power system. *Bulletin of Electrical Engineering and Informatics, 11*, 2442–2449. https://doi.org/10.11591/EEI.V11I5.3962
11. Falope, T. O., Lao, L., & Hanak, D. (2021). A unique three-step weather data approach in solar energy prediction using machine learning. Sustainable energy solutions for a post-COVID recovery towards a better future: Part VII, Bangkok: International Conference on Applied Energy; 2021, pp. 1–7. https://doi.org/10.46855/energy-proceedings-9863
12. Elma, O., Taşcıkaraoğlu, A., Tahir İnce, A., & Selamoğulları, U. S. (2017). Implementation of a dynamic energy management system using real time pricing and local renewable energy generation forecasts. *Energy, 134*, 206–220. https://doi.org/10.1016/J.ENERGY.2017.06.011

13. Minhas, D. M., Khalid, R. R., & Frey, G. (2017). Load control for supply-demand balancing under renewable energy forecasting. In *2017 IEEE 2nd international conference on direct current microgrids* (pp. 365–370). ICDCM. https://doi.org/10.1109/ICDCM.2017.8001071
14. Department for Energy Security and Net Zero. (2023). Greenhouse gas reporting: Conversion factors 2023. GOVUK 2023. https://www.gov.uk/government/publications/greenhouse-gas-reporting-conversion-factors-2023. Accessed 5 July 2023.

The Role of Integration of Hybrid Microgrids in Decarbonising Copper Deposits Mining in Australia

Hanrong Huang, Seher Ata, and Fiacre Rougieux

1 Introduction

The global energy landscape is undergoing a significant transformation, with a decisive shift towards renewable energy to mitigate climate change impacts and promote sustainable development. The mining sector, a cornerstone of Australia's economy, faces unique challenges in this transition, given its substantial energy consumption and environmental footprint [1–4]. This study delves into these challenges, focusing on copper deposit mining, a critical area due to its energy-intensive nature and the metal's pivotal role in clean energy applications [2, 5, 6]. This paper explores an avant-garde solution—the integration of hybrid microgrids within copper deposit mining operations in Australia. By synergising the realms of geological exploration and renewable energy systems, this research unveils the potential of solar, wind, and storage technologies harnessed through microgrids. These systems promise to revolutionise energy consumption patterns in mining, offering significant reductions in both cost and environmental footprint, thereby contributing to global decarbonisation efforts.

The research's novelty lies in its holistic approach, combining geological exploration with innovative energy solutions. Investigating copper deposits' distribution, depth, and ore grade, sets the stage for the strategic implementation of hybrid microgrids in mining operations. These microgrids, characterised by their

H. Huang (✉) · F. Rougieux
School of Photovoltaic and Renewable Energy Engineering, University of New South Wales, Sydney, Australia
e-mail: hanrong.huang@unsw.edu.au

S. Ata
School of Minerals and Energy Resources Engineering, University of New South Wales, Sydney, Australia

X. Wang (ed.), *Future Directions in Energy Engineering*, Green Energy and Technology, https://doi.org/10.1007/978-3-031-62042-3_4

adaptability and efficiency, represent a groundbreaking shift towards sustainable mining practices, harmonising economic growth with environmental stewardship.

2 Materials and Methods

The study's methodological framework integrated diverse data sources, including national geochemical databases [7], mining company disclosures, and renewable energy market analyses. The geological investigation mapped copper deposits nationwide, establishing a comprehensive understanding of their characteristics and potential for renewable energy integration.

Microgrid simulations were customised for each site, incorporating solar PV, wind energy, energy storage systems, and backup generators. The simulations, powered by advanced software tools HOMER [8–10], factored in local weather conditions, energy demand fluctuations, and potential grid instabilities, ensuring a realistic and reliable analysis.

Economic assessments were multifaceted, considering various cost components such as capital investment, operational expenses, fuel costs, and decommissioning. The Levelised Cost of Electricity (LCOE), a critical metric, was calculated for each microgrid configuration, providing a clear economic performance indicator. In the calculation of LCOE, the following formula was used, assuming a discount rate of 8%, and 25 years of project lifetime.

$$\text{Levelised Cost of Electricity} = \frac{\sum_{t=0}^{n} \frac{I_t + M_t}{1 + r^t}}{\sum_{t=0}^{n} \frac{E_t}{1 + r^t}} \tag{1}$$

Where:

I_t = Investment costs in the year t.
M_t = O&M costs in the year t.
E_t = Fuel costs in the year t.
r = Discount rate.
n = Project lifetime.

These assessments were crucial in benchmarking microgrid solutions against conventional diesel-powered generators and existing electricity grids, highlighting the competitive advantages and limitations of renewable alternatives.

3 Results and Discussion

The findings from Figs. 1 and 2 presented a multifaceted view of copper mining's potential transition to renewable energy. The copper deposits, scattered across Australia, exhibited significant variation in ore grades and depth, directly influencing the microgrids' design and economic viability. Copper deposits with a higher ore grade than the cut-off grade are expected to have a higher economic payback for ore extraction [11]. As depicted in Fig. 1, regions such as Queensland (QLD), South Australia (SA), and Western Australia (WA) are identified as primary locales for high-grade copper deposits. Another critical determinant in assessing the economic feasibility of mining operations pertains to the depth of the deposit. Deposits located near the surface offer economic benefits, making them well-suited for surface mining techniques like open-pit mining. On the contrary, subterranean deposits necessitate more financially onerous underground mining techniques. Generally, copper deposits with a depth of less than 300 m are deemed favourable for open-pit mining operations [12]. It can be found in Fig. 2 that a significant number of copper deposits are less than 300 m in depth and have a fairly high ore grade. These specific deposits are identified as propitious zones for renewable energy investments, heralding a reduction in mining operational expenditures and carbon emissions, whilst augmenting economic returns.

Figures 3 and 4 present the simulated outcomes of hybrid microgrid configurations established at each copper deposit. When compared to the Levelised Cost of Electricity (LCOE) of the diesel generator baseline case (0.42 \$/kWh) derived from the simulation, the LCOE of the hybrid microgrids is observed to be lower at a multitude of copper deposits. This disparity is likely attributed to the elevated renewable fraction inherent in the hybrid microgrids, as illustrated in Fig. 4, wherein a notable correlation between LCOE and renewable fraction is observed. The LCOE of hybrid microgrids demonstrates a declining trend in conjunction with an increase

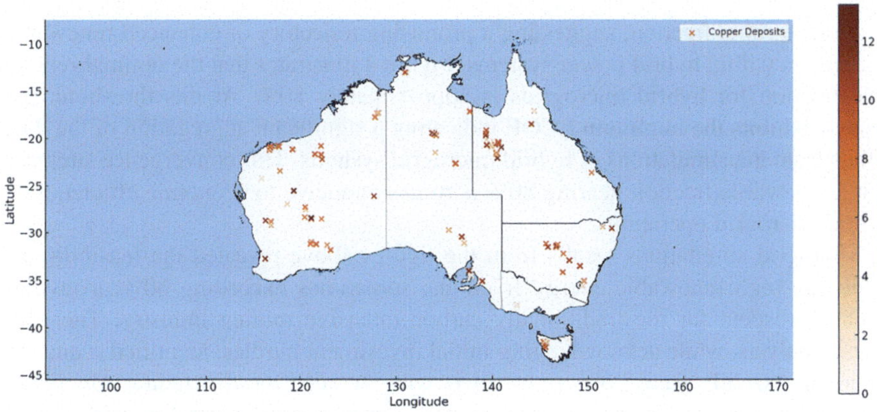

Fig. 1 Copper deposits geographic distribution and their associated ore grade in Australia

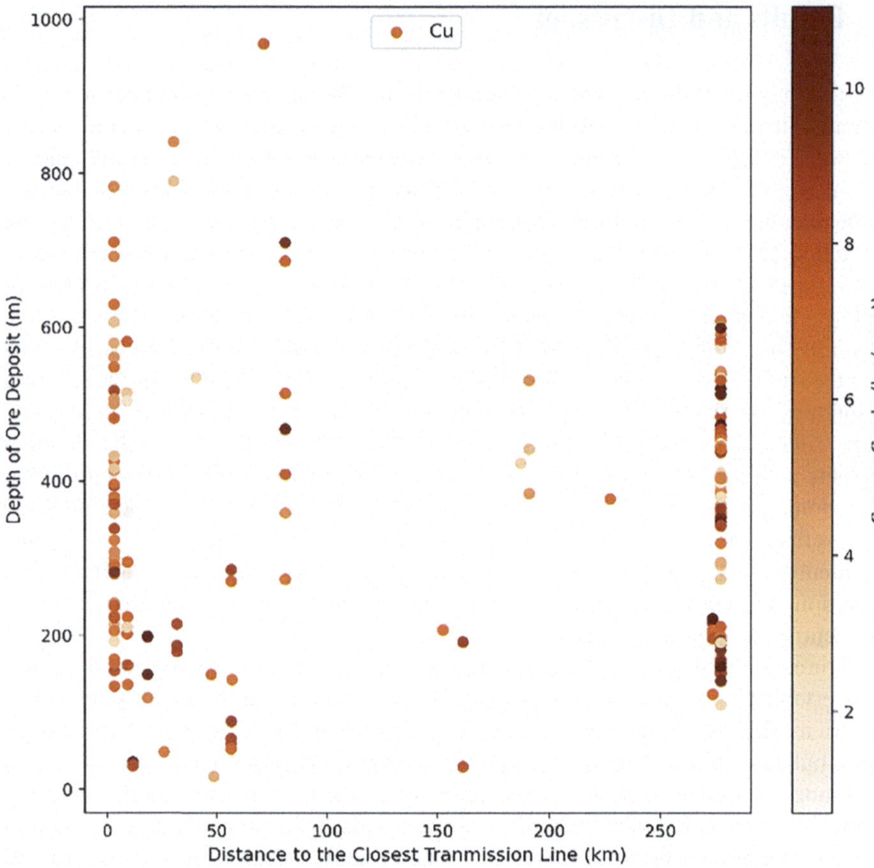

Fig. 2 Copper deposits distribution grouped by their distance to the closest transmission line and the depth of ore deposits. The colour of scatters shows the ore grade of copper deposits

in the renewable fraction, suggesting a promising trajectory of enhanced renewable integration within hybrid power systems. Figure 4 illustrates that the optimal renewable fraction for hybrid microgrids is approximately 80%. At this threshold, the system exhibits the minimum LCOE indicating a significant aggregation of the data points from the simulations of hybrid microgrid systems. This convergence suggests that a renewable fraction nearing 80% is most conducive to economic efficiency in hybrid microgrid operations.

Microgrid simulations results from the figures above revealed the feasibility of achieving high renewable energy fractions, sometimes exceeding 80%, a revolutionary prospect for the traditionally carbon-intensive mining industry. The economic analysis, while acknowledging initial investment hurdles, identified scenarios where microgrids were cost-competitive with or superior to, traditional energy sources. This economic promise, however, is contingent on regional policies, financial incentives, and societal perceptions of renewable energy.

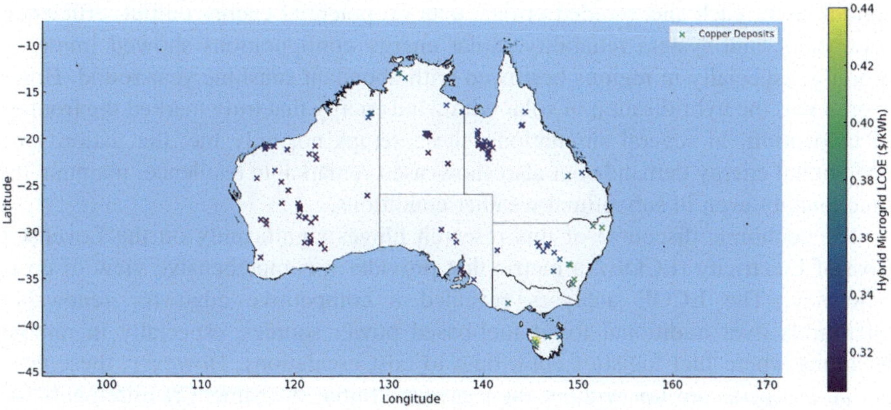

Fig. 3 The LCOE of hybrid microgrid simulated at each copper deposit

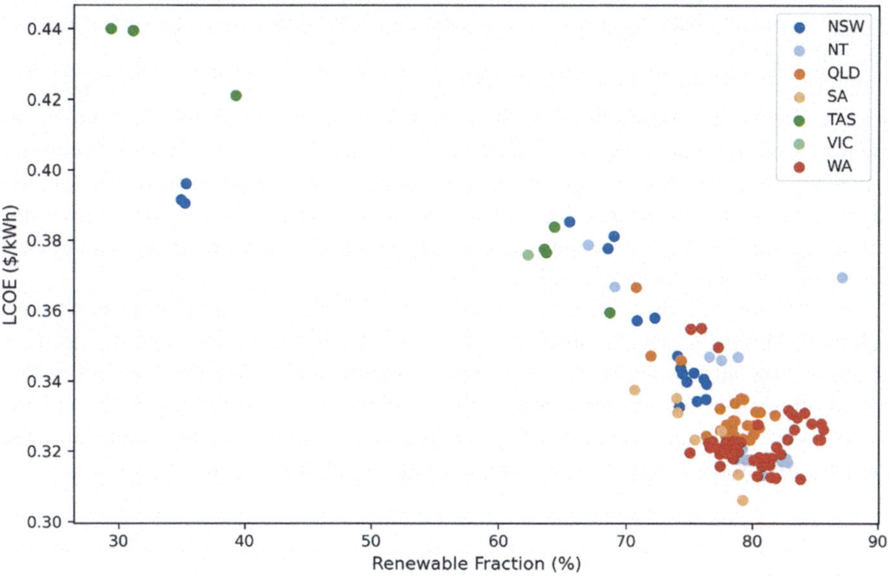

Fig. 4 The renewable fraction versus the LCOE of hybrid microgrids, grouped by states in Australia

The comparative analysis also underscored the indirect benefits of microgrids, such as enhanced energy security, reduced fuel price volatility, and positive environmental and social impacts. These factors, often overlooked in traditional cost assessments, could be pivotal in accelerating the adoption of renewable energy in mining.

In the realm of microgrid performance, the research ventured deep into the analytics of energy. Simulations, tailored to the unique topographical and climatic conditions of each site, yielded critical data on potential energy output, efficiency parameters, and system reliability. Solar energy configurations showed immense promise, especially in regions bestowed with abundant sunshine year-round. However, it was the hybridisation of solar with wind energy that truly marked the frontier of innovation. In several simulations, these setups not only met the majority of operational energy demands but also showcased remarkable resilience, maintaining functionality even in suboptimal weather conditions.

The economic discourse of this research hinges significantly on the Levelised Cost of Electricity (LCOE), a metric that provides a comprehensive view of cost-efficiency. The LCOE analyses revealed a competitive edge for renewable microgrids over traditional fossil fuel-based power sources, especially in remote locations where fuel logistics contribute to cost escalations. However, these economic prospects are not without their caveats. Initial investment requirements for renewable setups are substantial, posing financial challenges, particularly for smaller mining operations. Yet, these hurdles are not insurmountable. With targeted government subsidies, tax incentives for renewable energy, and financing models that mitigate upfront costs, the economic scales could decidedly tip in favour of renewable microgrids.

Confronting the complexities of large-scale implementation, the research acknowledges the multitude of challenges that lie ahead. Technological advancements, particularly in energy storage and transmission, are paramount to overcoming efficiency bottlenecks. Financial and regulatory landscapes, currently geared towards conventional energy practices, require significant restructuring. This transition, monumental in its scope, calls for cohesive strategies, uniting stakeholders across government, industry, and civil society.

In reflection, the discussions within this study highlight a horizon imbued with potential. The path forward, albeit steeped in challenges, glimmers with the promise of innovation, sustainability, and profound economic and social transformation. The metamorphosis of the mining sector, with hybrid microgrids at its vanguard, beckons as both a necessity and an opportunity—a chance to redefine paradigms and steer the industry towards a future that harmonises with the global sustainability ethos.

4 Conclusions

The research underscores the transformative potential of hybrid microgrids with high renewable penetration in the mining sector, a shift that could redefine energy consumption patterns and environmental impacts in one of Australia's key industries. The findings of this investigation indicate that hybrid microgrids, comprising roughly 80% renewable energy fraction, possess considerable potential to reduce the LCOE as well as the carbon emissions associated with the mining of copper deposits throughout Australia.

While the path to widespread adoption is complex, involving economic, policy, and social factors, the benefits are indisputable. Reduced greenhouse gas emissions, cost savings, and improved energy resilience stand out as compelling incentives.

Future studies should extend this research scope, exploring diverse financing strategies, policy mechanisms, and community engagement practices that could facilitate renewable energy transition in mining. Additionally, expanding the focus to include other minerals and global regions will enrich the understanding and applicability of renewable microgrids in mining worldwide.

References

1. Alves, W., Ferreira, P., Araújo, M., & Cézar, J. (2020, June). Renewable energy for sustainable mining. *International Journal for Quality Research, 14*(2), 593–600. https://doi.org/10.24874/IJQR14.02-16
2. Behar, O., Sbarbaro, D., & Moran, L. (2021, March). Which is the most competitive solar power technology for integration into the existing copper mining plants: Photovoltaic (PV), Concentrating Solar Power (CSP), or hybrid PV-CSP? *Journal of Cleaner Production, 287*, 125455. https://doi.org/10.1016/j.jclepro.2020.125455
3. Alova, G. (2018). *Integrating renewables in mining: Review of business models and policy implications* (p. 49). https://doi.org/10.1787/5bbcdeac-en
4. *Decarbonizing the mining industry: Achieving 2030 and 2050 goals – KPMG Canada*. KPMG. Accessed 12 Apr 2023. [Online]. Available: https://kpmg.com/ca/en/home/insights/2021/11/decarbonizing-the-mining-industry.html
5. Behar, O., et al. (2021, October). The use of solar energy in the copper mining processes: A comprehensive review. *Cleaner Engineering and Technology, 4*, 100259. https://doi.org/10.1016/j.clet.2021.100259
6. Gómez, J. S., et al. (2020). An overview of microgrids challenges in the mining industry. *IEEE Access, 8*, 191378–191393. https://doi.org/10.1109/ACCESS.2020.3032281
7. Australia Government-Geoscience. *Critical minerals at geoscience Australia*. Applying geoscience to Australia's most important challenges. [Online]. Available: https://www.ga.gov.au/scientific-topics/minerals/critical-minerals
8. Santos, L. H. S., Silva, J. A. A., López, J. C., Arias, N. B., Rider, M. J., & Da Silva, L. C. P. (2021, September). Integrated optimal sizing and dispatch strategy for microgrids using HOMER Pro. In *2021 IEEE PES innovative smart grid technologies conference – Latin America (ISGT Latin America)* (pp. 1–5). https://doi.org/10.1109/ISGTLatinAmerica52371.2021.9543015
9. H. Bitaraf and B. Buchholz 2018 Reducing energy costs and environmental impacts of off-grid mines .
10. Castro, G. A., Murkowska, M. I., Rey, P. Z., & Anvari-Moghaddam, A. (2020, October). Operational planning of a hybrid power plant for off-grid mining site: A risk-constrained optimization approach. In *IECON 2020 the 46th annual conference of the IEEE Industrial Electronics Society* (pp. 4587–4592). https://doi.org/10.1109/IECON43393.2020.9254402
11. Beck, D. (2013). Determination of optimal mining cut-off grades: Mathematical formulation and solution algorithm: (Case STUDY: Hinoba Copper Mine in the Philippines). https://doi.org/10.13140/RG.2.2.35292.80007.
12. Special Report. *EXPLAINER: How drilling results make the grade—copper and zinc*. Stockhead. Accessed 23 Oct 2023. [Online]. Available: https://stockhead.com.au/resources/how-do-drilling-results-make-the-grade-copper-and-zinc/

Utilizing Wind Farms and Electric Vehicles to Mitigate the First and Second Frequency Dips in Power Systems

Seyed Amir Hosseini and Saeed Peyghami

1 Introduction

Employing renewable energy sources (RESs) like wind and solar in the modern power system has significantly increased in recent years [1]. These resources besides HVDC links decrease the power system inertia dramatically, making the system more vulnerable to unexpected load changes and failures of generation units or transmission lines. As a result, keeping the power system frequency within a reasonable range is a demanding task [2, 3].

Even though wind turbines are separated from the power system frequency via power converters, effective control schemes allow them to collaborate in inertia emulation. Brisebois and Aubut [4] introduce a highly effective approach for enhancing power system frequency using wind turbines. In this source, the Hydro-Quebec electricity transmission system proposes a method for emulating inertia to counteract the impact of reduced system inertia. In the face of large generation unit outages, the power system frequency starts decreasing. At the first step, a wind turbine temporarily raises its output power partly eliminating the first frequency dip. After finishing the support state, the wind turbine's recovery state will be started and causes a new frequency dip called the second frequency dip. The system operators are not in favour of this second dip because this dip can be deeper in comparison to the first frequency dip [5]. Modern power systems contain a high share of wind farms. More support causes a deeper second frequency dip. Therefore, the first and second frequency dips should be eliminated somehow.

Over the past few years, there has been a growth in the number of EVs on the road that can be a remarkable option to collaborate in the power system frequency regulation [6].

S. A. Hosseini (✉) · S. Peyghami
Aalborg University, Aalborg, Denmark
e-mail: Saho@energy.aau.dk; Sap@energy.aau.dk

X. Wang (ed.), *Future Directions in Energy Engineering*, Green Energy and Technology, https://doi.org/10.1007/978-3-031-62042-3_5

2 EV Participation in Frequency Control

2.1 Proposed Virtual Inertia Controller for EVs

To use EVs' frequency regulating capacity, a virtual inertia controller is proposed. The EV current is meticulously controlled in increments of 1A, ranging from 6A to 16A, in compliance with the IEC 61851 standard [7]. The nominal current for EVs is set at 11A.

EVs plug into a charging station for 90% of the times [8]. They can emulate the behaviour of conventional power plants by supplying active power that is directly proportional to the rate of change of frequency (ROCOF). The proposed virtual inertia controller scheme is shown in Fig. 1. The theoretical and actual output current responses are demonstrated by red and blue curves, respectively. Here, ROCOF is measured continuously, and based on the data depicted in Fig. 1, the EV's current is adjusted accordingly. Thanks to the narrow dead band (±0.3 Hz/s) in Fig. 1, EVs are spared from participating in frequency regulation for minor frequency deviations, ensuring the preservation of their battery lifespan.

A thorough model for the EV's involvement in frequency regulation is depicted in Fig. 2. It consists of an EV dynamic model and a virtual inertia controller. In the EV dynamic model, $D_{\text{A\&C}}$ represents inverter activation and communication latency, whereas V and P represent the EV's voltage and active power, respectively. In the virtual inertia control model, the curve is based on IEC 61851 standard, *f* is the frequency of the power system and $D_{\text{M\&C}}$ is measurement and communication latency.

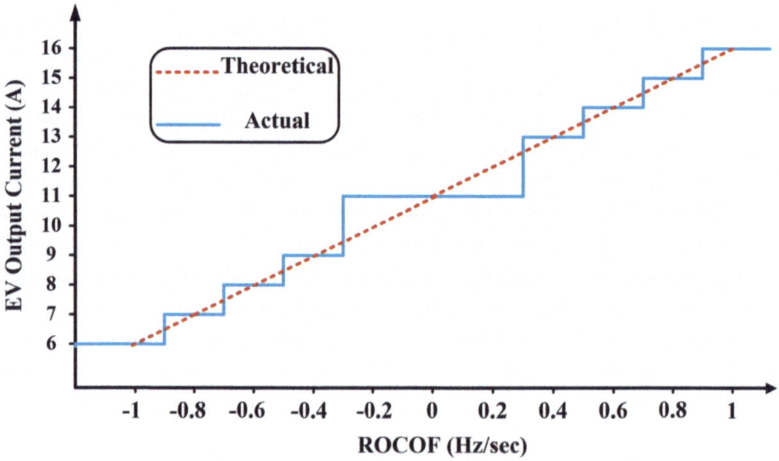

Fig. 1 Droop characteristic of virtual inertia controller of EV

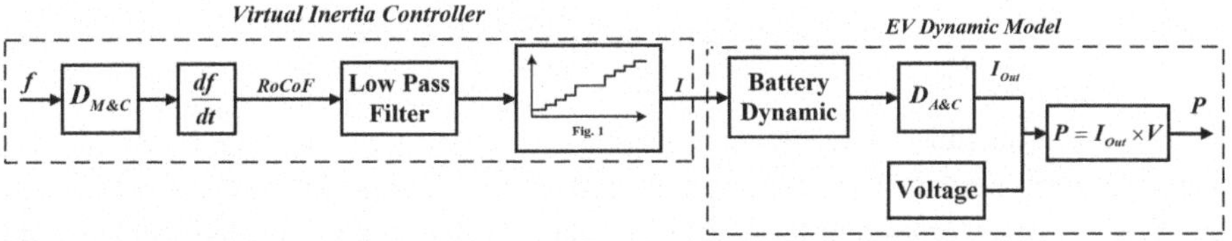

Fig. 2 EV dynamic model and virtual inertia controller diagram

2.2 *Sequencing EVs Parking Lots' Participation in Power System Frequency Control*

The sequential participation sequence of various electric vehicle parking lots (EVPLs) in the control of the power system's frequency is another crucial issue that should be addressed. To cover this concern, a threshold value $\Delta f_T^{\,i}$ is dedicated to the *i*th EVPL. The first EVPL joins the efforts in frequency control immediately upon the power system's frequency deviation exceeding the minimum frequency violation threshold, denoted as $\Delta f_{\min}$.

Moreover, when the power system's frequency deviation surpasses another predefined threshold, denoted as $\Delta f_{\max}$, all EVPLs collaborate collectively to regulate the system's frequency. The area situated between these established thresholds is subdivided into segments, each representing a group of potential EVPLs, and the threshold value $\Delta f_T^{\,i}$ for each EVPL is determined accordingly. The turn of EVPLs collaboration rotates hourly to evenly spread the burden on the EVPLs. EVPLs with smaller Δf_T participate in the frequency regulating process earlier than others.

3 Wind Farms Collaboration in Frequency Control

3.1 *Principles of Wind Turbines*

The single mass model for a wind turbine is explained as follows:

$$J\frac{d\omega}{dt} = T_{\text{aero}} - T_{\text{gen}} = \frac{\frac{\rho}{2}\pi R^2 v^3 C_p(\lambda,\beta)}{\omega} - \frac{\rho}{2}\pi R^2 C_p(\lambda^*)\left(\frac{R}{\lambda^*}\right)^3 \omega^2 \tag{1}$$

In (1), J is wind turbine hub inertia (kg.m^2), T_{aero} and T_{gen} are aerodynamic, and generator torques of a wind turbine. The wind turbine's operating point is modified by the generator's torque. At the steady state, λ^* is adjusted based on the wind velocity. ΔT is connected to the generator torque in accordance with the following equation to create the inertia response of the wind turbine.

$$T_{\text{gen}} = \frac{\rho}{2}\pi R^2 C_p(\lambda^*)\left(\frac{R}{\lambda^*}\right)^3 \omega^2 + \Delta T \tag{2}$$

Various controllers have been introduced to facilitate the release of inertial response from wind turbines, aiding in frequency regulation. One of these controllers involves a power reference term. In this scenario, the change in generator torque (ΔT) is as follows where M is a constant.

$$\Delta T = \frac{M}{\omega} \tag{3}$$

By modifying the value of M, this controller alters the speed of the turbine hub, consequently affecting the amount of kinetic energy released by the wind turbine. The torque curves relative to the hub speed of the wind turbine are illustrated in Fig. 3, with T_{gen} and T'_{gen} representing the steady-state generator torques of the wind turbine before and after the application of ΔT. These torque values are defined by the below equation:

$$\begin{cases} T_{\text{gen}} = \frac{\rho}{2}\pi R^2 C_p(\lambda^*)\left(\frac{R}{\lambda^*}\right)^3 \omega^2 \\ T'_{\text{gen}} = \frac{\rho}{2}\pi R^2 C_p(\lambda^*)\left(\frac{R}{\lambda^*}\right)^3 \omega^2 + \frac{M}{\omega} \end{cases} \tag{4}$$

As depicted in Fig. 3, when the generator torque is increased (indicated by a positive M value), it results in a decrease in hub speed. Additionally, the red curve

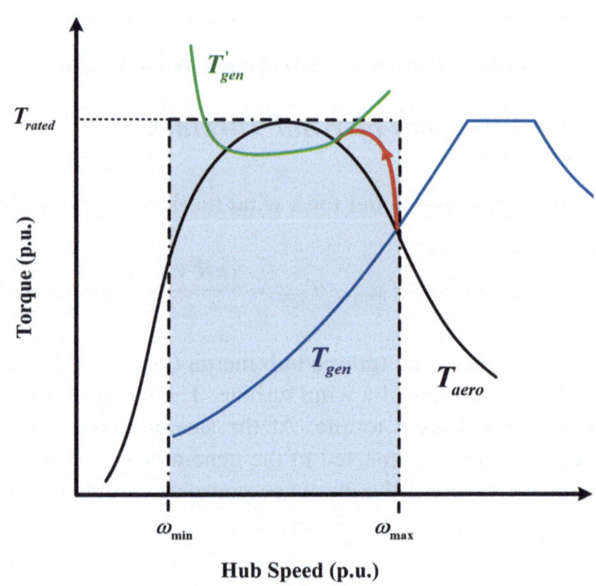

Fig. 3 Torques related to a wind turbine

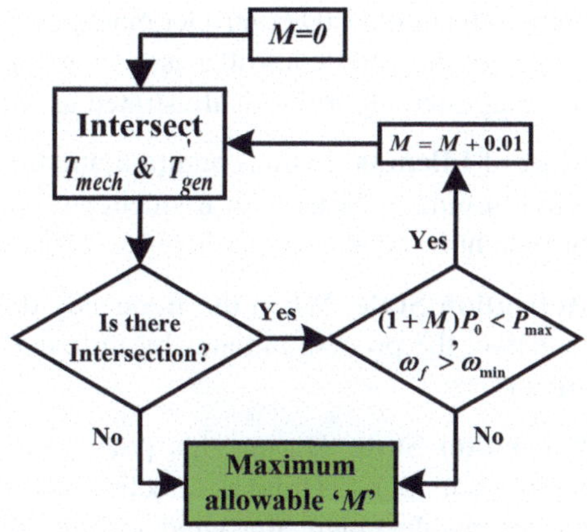

Fig. 4 Suggested algorithm for determining the highest attainable value of M

illustrates the transient path. It is important to emphasize that for stable operation, T'_{gen} should intersect with T_{aero}. Furthermore, after boosting the power of the wind turbine, the generated output power must remain below the wind turbine's maximum capacity, and the final hub speed should be higher than the minimum hub speed. Figure 4 shows a technique for determining the maximum permissible value for M. Here, P_0 represents the initial power of the wind turbine before any increase, P_{max} stands for the maximum achievable output power of the wind turbine, and ω_f represents the final hub speed after the power increment. In this algorithm, the initial assumption is that $M = 0$, and the curves for T_{gen} and T'_{gen} intersect. If, after increasing M by 0.01 p.u., the updated wind turbine output power remains below P_{max}, and the hub speed at the intersection point (ω) exceeds the minimum allowable hub speed value (ω_{min}), the process continues. The intersection between T_{gen} and T'_{gen} curves is reevaluated, and this iteration repeats. The process terminates when there is no longer an intersection between T_{gen} and T'_{gen} curves, or when the wind turbine output power surpasses P_{max}, or when ω falls below ω_{min}. At this point, the new value of M is identified as the maximum allowable increment in the wind turbine's output power.

3.2 *Suggested Control Scheme*

The initial concept of the proposed controller draws inspiration from Hydro-Quebec's inertia requirement, which aims to emulate inertia from renewable energy sources. In Hydro-Quebec's proposed scheme, there are no maximum and minimum boundaries for power increment and recovery duration, respectively. Considering the dynamics of the wind turbine, it may not be feasible to enhance its power output by more than 6% and have it fully restored within a time frame of less than 20 s. To address this challenge, an algorithm has been put forward. This algorithm comprises various states that are activated based on the measured frequency and its ROCOF value. The initial PID controller manages the ramping up and uniform support states, while another PID controller is responsible for the recovery state. The structure of the suggested algorithm is illustrated as follows:

State of Idleness In this condition, the frequency deviation is insufficient to ask the wind turbine to assist with frequency regulation. The operational point of the wind turbine hub speed is found here by intersecting the T_{aero} and T_{gen} curves.

Activation State When the frequency drops below 49.95 Hz and the ROCOF is negative, the process initiates by utilizing ΔT from Eqs. (3) and (4) as the control signal.

Escalation State During this phase, the wind turbine's output power increases rapidly within 1 s. To be as effective as possible in stabilizing the frequency within an acceptable range, the wind turbine aims to maximize its output power. The maximum allowable value for M, determined based on wind velocity as shown in Fig. 4, plays a crucial role in this process. If the time elapsed since activation is less than 1 s and the frequency exceeds 49.95 Hz, the turbine ceases its support and transitions to the recovery state. However, if the time reaches 1 s and the frequency remains at or below 49.95 Hz, the subsequent state is then pursued.

Uniform Support State In this step, the wind turbine provides a constant increment of power, which is set to the maximum value M. The wind turbine remains in this state for a maximum duration of 10 s or until the wind turbine's hub speed (ω) matches the minimum acceptable hub speed (ω_{min}). The first PID controller is intended to respond to escalation and uniform support states. In this controller, the control signal is determined by ΔT from Eq. (3), and it undergoes a two-phase increase. Consequently, the wind turbine's hub speed decreases, as indicated in Fig. 3.

Recovery State Here, the wind turbine's hub speed is restored using a second PID controller. This controller aims to reduce ΔT to zero, resulting in the wind turbine's hub speed being adjusted by intersecting the T_{aero} and T_{gen} curves at a new wind velocity.

The suggested algorithm's block diagram is given in Fig. 5, where AT is the activation time and USD represents uniform support duration and activation time.

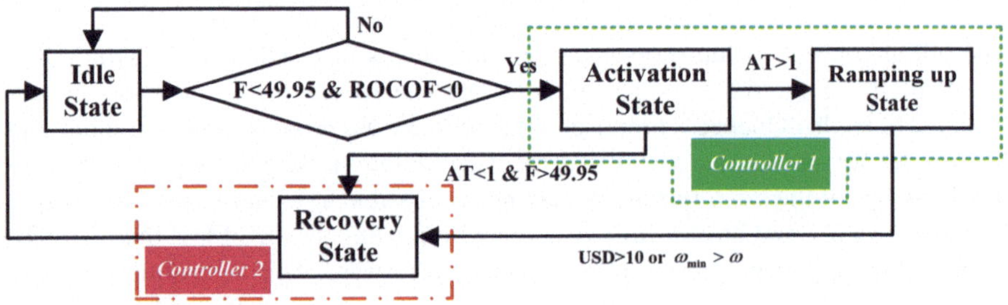

Fig. 5 Suggested control scheme

3.3 Alleviating the Second Frequency Dip

To mitigate the second frequency dip, it is necessary to have slow transitions and recovery states for the wind farms, which can take a significant amount of time. However, as this duration increases, the recovery of the power system frequency and the wind turbines' output power may become unacceptably long. One potential solution to reduce the impact of the second frequency dip is to employ electric vehicles. The challenge here lies in determining the optimal moment to activate the EVs. Essentially, when the second frequency dip occurs, the EV controller must be triggered. In situations where the wind farms do not participate in power system frequency regulation as wind velocity increases, the wind turbine's hub speed (ω) rises, leading to an increment in the wind turbine's output power (P_{wt}), and vice versa. In under-rated wind velocities, the following equation holds true:

$$\left(\frac{d\omega}{dt}\right) \times \left(\frac{dP_{\mathrm{wt}}}{dt}\right) > 0 \tag{5}$$

When the wind turbine actively participates in frequency regulation, the wind turbine's output power (P_{wt}) increases while the hub speed (ω) decreases during the escalation and uniform support states. Conversely, in the recovery state, P_{wt} decreases, and ω rises. Consequently, it can be inferred that when the hub speed increases, $(d\omega/dt) > 0$, and the P_{wt} decreases, $(dP_{\mathrm{wt}}/dt) < 0$, the second frequency dip occurs. Therefore, when both conditions, $(d\omega/dt) > 0$ and $(dP_{\mathrm{wt}}/dt) < 0$, are met, it is advisable to activate the EVs controller to mitigate the extent of the second frequency dip.

4 Simulation Results

To assess and contrast the effectiveness of the suggested controllers and algorithms, certain examinations and simulations have been conducted using the adapted IEEE 39-bus test power system. The details regarding the power system's specifications, EVs, and wind turbine characteristics can be found in [9]. The performance assessment comprises four distinct states and in all of them wind farms are responsible for supplying approximately 40% of the total power and at $t = 5$ s, a step load increase of $\Delta P = 200$ MW is introduced into the system as a disturbance.

At the first state, the power system is assumed to be equipped solely with droop speed governors along with AGC. The frequency excursion is depicted in Fig. 6 by the solid blue curve, indicating a notable frequency deviation. Consequently, in the second state, the introduced controller in Fig. 5 for enabling wind turbine collaboration in frequency regulation is implemented. The nadir frequency improved significantly as depicted in Fig. 6 by the dashed black curve. In the third state, 10,000 EVs with the proposed virtual inertia controller are introduced to the power

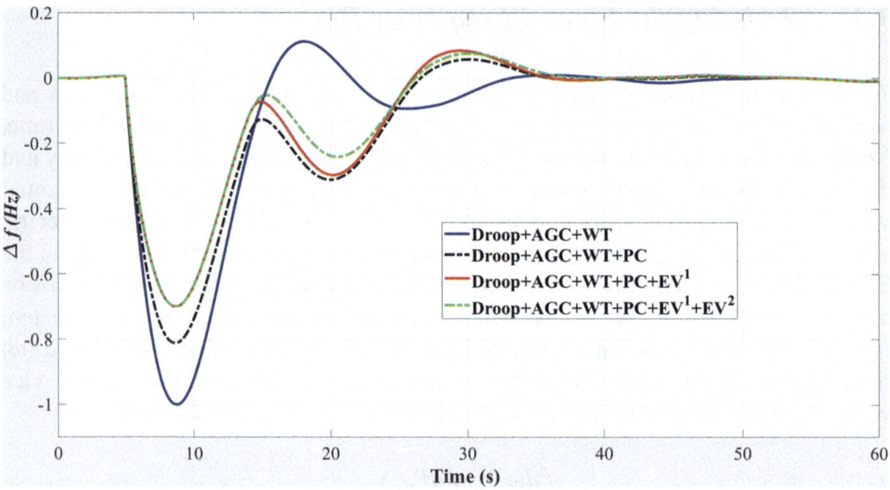

Fig. 6 Power system frequency behaviour at different states

system. As can be seen by the solid red curve in frequency curves, there is a notable reduction demonstrating the effectiveness of EVs and WT controller. The only drawback is the considerable second frequency dip. To cover this concern in the fourth state, the suggested method for alleviating the second frequency dip is adopted. When specific conditions regarding $d\omega/dt$ and dP_{wt}/dt are met ($d\omega/dt > 0$ and $dP_{wt}/dt < 0$), the second frequency dip is going to happen, and 6000 other EVs are added to the third state to reduce the magnitude of the second frequency dip. The frequency excursion of this state is shown by the dashed green curve in Fig. 6. Notably, the second frequency dip similar to the first frequency dip is reduced significantly.

5 Conclusion

In this study, the objective was to enhance the power system's ability to respond to frequency fluctuations when wind power plants are integrated. This was achieved by introducing EVs and wind turbines into the power grid. Conforming to the IEC 61851 standard, a virtual inertia controller was proposed for EV collaboration in regulating power system frequency. A dedicated algorithm was devised to leverage the frequency-regulating capabilities of wind turbines. This algorithm enables wind turbines to rapidly adjust their power output to the maximum allowable value in response to disturbances and then maintain this level. Additionally, a separate algorithm was presented to determine the maximum allowable power increment. When wind turbines operate in a consistent support state for 10 s or when the hub speed falls below the minimum acceptable value, the wind turbines are restored to their normal state. Two PID controllers were specifically designed to oversee the support and recovery phases of wind turbine operation. These proposed algorithms were applied to the modified IEEE 39-bus test power system, and the outcomes verified the effectiveness of the suggested methodologies. Notably, incorporating EVs and wind turbines into the power system independently improved frequency response. However, the most efficient enhancement in system frequency behaviour was observed when both the EVs and the proposed wind turbine algorithm were deployed simultaneously.

References

1. Hosseini, S. A., Toulabi, M., Ashouri-Zadeh, A., & Ranjbar, A. M. (2019). Modified power reserve management solution in power system considering frequency constraints. *IEEE Systems Journal, 14*(1), 1125–1134.
2. Muljadi, E., Gevorgian, V., Singh, M., & Santoso, S. (2012). Understanding inertial and frequency response of wind power plants. *IEEE*, 1–8.
3. Hosseini, S. A., Fotuhi-Firuzabad, M., Dehghanian, P., & Lehtonen, M. (2024). Coordinating demand response and wind turbine controls for alleviating the first and second frequency dips in wind-integrated power grids. *IEEE Transactions on Industrial Informatics, 20*, 2223.
4. Brisebois, J., & Aubut, N. (2011). Wind farm inertia emulation to fulfill Hydro-Québec's specific need. In *2011 IEEE power and energy society general meeting* (pp. 1–7). IEEE.
5. Kang, M., Lee, J., & Kang, Y. (2015). Modified stepwise inertial control using the mechanical input and electrical output curves of a doubly fed induction generator. In *2015 9th international conference on power electronics and ECCE Asia (ICPE-ECCE Asia)* (pp. 357–361). IEEE.
6. Tomic, J., & Kempton, W. (2007). Using fleets of electric-drive vehicles for grid support. *Journal of Power Sources, 168*, 459–468.
7. I. E. Commission. (2017). *Electric vehicle conductive charging system-part 1: General requirements*. IEC Standard 61851.
8. Rahman, T., & Qu, Z. (2017). The role of electric vehicles for frequency regulation during grid restoration. In *2017 IEEE power & energy society general meeting* (pp. 1–5). IEEE.
9. Homepage. https://www.dropbox.com/s/3upke27u2bi6ssz/Data.xlsx?dl=0

Performance Analysis of PEMFC Coupled with Recuperative Organic Rankine Cycle for Waste Heat Recovery

Pratyusha Padhi and Murugan Sivalingam

1 Introduction

Fuel cells convert the chemical energy in hydrogen or hydrogen-based fuel into electricity by an electrochemical conversion process [1]. Among the different types of fuel cell, the proton exchange membrane fuel cell (PEMFC) has received profound interest due to its added advantages like longer operating life, low operating temperature, and higher stack power density compared to its competitors. A low-temperature PEMFC stack ideally has an operating temperature range of 60–100 °C with an electrical conversion efficiency of 30–50%. At this operating temperature range a huge portion of the hydrogen energy entering the fuel cell is wasted in the form of heat, generated as a by-product of the fuel cell reactions. This low-grade heat, at low operating temperatures, has huge potential to be utilized independently for various onsite thermal applications because of its affluence. The amount of waste heat generated under normal working conditions is almost equal to the output power. In recent years, the recovery waste heat from PEMFCs and utilization has gained significant interest among researchers and scientists. The waste heat was used to produce mechanical power electrical power, cooling, energy and fuels [2]. Among the waste heat recovery techniques, the organic Rankine cycle (ORC) is a widely used system for recovering heat from PEMFC cogeneration or trigeneration systems [3–5]. Attempts were made to improve efficiency by using a recuperator which used the heat from the refrigerant coming out of the turbine in the gaseous phase to increase the temperature of the liquid refrigerant coming out of the pump. Sheshpoli et al. investigated a hybrid system consisting of 1180 kW low-temperature PEMFC coupled with VCR cycle using RORC [6]. A comparison was made between two working fluids, viz. R-134a and R-245fa. Kaushik et al.

P. Padhi · M. Sivalingam (✉)
Department of Mechanical Engineering, National Institute of Technology, Rourkela, India
e-mail: murugans@nitrkl.ac.in

X. Wang (ed.), *Future Directions in Energy Engineering*, Green Energy and Technology, https://doi.org/10.1007/978-3-031-62042-3_6

coupled a RORC with a hydrogen-fuelled PEMFC stack and analysed the hybrid system performance [7]. Three refrigerants, namely R-245fa, R-123, R-245ca were used to evaluate the hybrid system performance. R-245fa working fluid offers the optimum performance and efficiency for extracting the maximum possible heat from the fuel cell for better functioning of the overall system with minimal losses. About 2% gain in electrical efficiency is obtained by coupling RORC with an electrical efficiency of 32.10%.

This research work evaluated a few performance parameters of a PEMFC coupled with an RORC system. The heat generated in the stack was channelized through two heat exchangers to reach the evaporator which required a sufficient amount of heat to produce useful work. For simplification, the additional work input for operating auxiliary components in the PEMFC system was neglected. A thermodynamic modelling approach was followed to evaluate a few performance parameters of the proposed hybrid system considering different operating conditions like the operating temperature of the fuel cell, the refrigerants used in the RORC cycle and the mass flow rate of the RORC refrigerants. The results obtained from the MATLAB simulation were theoretically analysed to predict the effect of operating parameters on fuel cell performance and presented in this paper.

2 System Description

2.1 Recuperative Organic Rankine Cycle and Proposed Hybrid System

Figure 1 shows a schematic diagram of the Recuperative Organic Rankine Cycle (RORC) system. Heat from the PEMFC stack is exchanged with the RORC liquid refrigerant stream entering the evaporator at point 1. This heat is used to increase the temperature and change the phase of the working fluid to a vapor state while exiting the evaporator at point 2. The vapor stream of working fluid at point 2 enters the turbine where the enthalpy of the stream drops and exits the turbine at point 3. At this point, the vapor is in a superheated state. The process inside the turbine is not reversible adiabatic. The working fluid stream exchanges some heat with the gaseous refrigerant coming from the turbine inside the recuperator and finally leaves the recuperator at point 4 and enters the condenser. The glycol-water mixture acts as a secondary refrigerant and takes away the heat rejected by the gaseous refrigerant in the condenser. The refrigerant leaving the condenser at point 5 is in a saturated state. This stream is pumped to point 6 and enters the recuperator at point 6 to absorb the heat from the stream at point 3. The stream exiting the recuperator enters the evaporator at point 1 and the cycle begins once again. The recuperator increases the temperature of the stream entering the evaporator without adding additional heat and hence increases the overall efficiency of the cycle.

Fig. 1 Recuperator in ORC

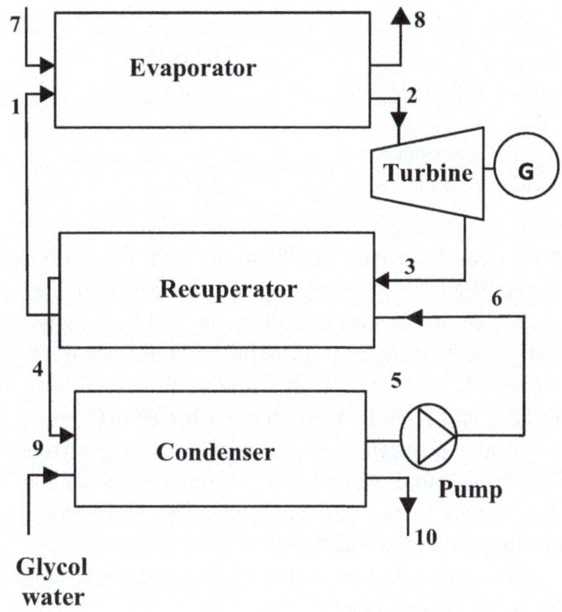

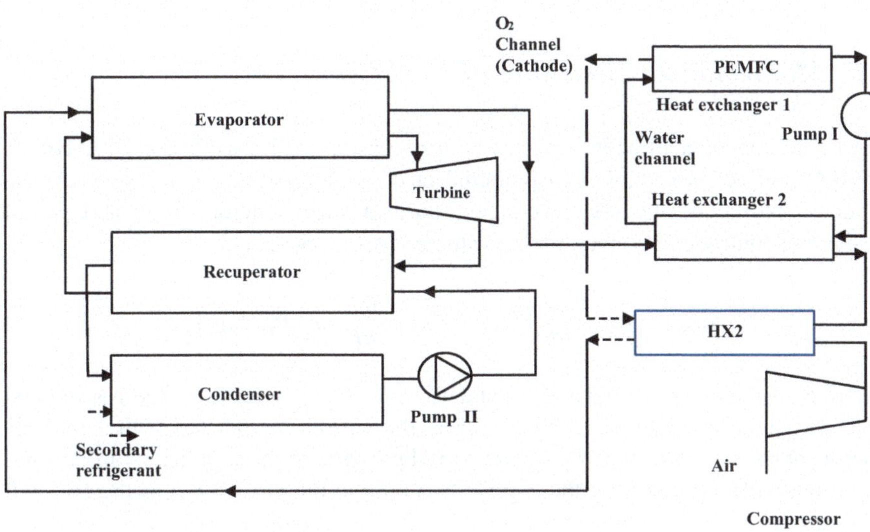

Fig. 2 Proposed hybrid system

Figure 2 portrays the arrangement of the proposed hybrid system in which recuperator and ORC (RORC) are integrated with PEMFC. The air entering the PEMFC stack is compressed by the compressor to its operating pressure and finally cools down to the operating temperature by rejecting heat from the cooling water stream inside the heat exchanger 2 before entering the PEMFC stack at the cathode.

Table 1 Important physical properties of the working fluids [8]

Property	R245fa	R123	R-141b
Chemical formula	$CF_3CH_2CHF_2$	CF_3CHCl_2	CH_3CCl_2F
Molecular weight	134	156	116.95
Critical temperature (°C)	154	184	204.15
Critical pressure, MPa	3.65	3.67	4.25

Moreover, hydrogen fuel from the hydrogen delivery system reaches the anode side of the PEMFC. The reaction of hydrogen with air (oxygen) inside the PEM fuel cell generates power along with water and heat as by-products. Cooling water absorbs the waste heat from the PEMFC and converts it into usable form inside the turbine of the RORC system. In this study, three working refrigerants including R-245fa, R-123, and R-141b were chosen for RORC and their performance was examined. Important properties of the three working refrigerants are given in Table 1. The thermodynamic analysis was assumed to be in a steady state with no heat loss from the system to the surroundings. The temperature at the inlet and outlet of each component is assumed to be constant.

The pressure drop in the pipe, condenser, recuperator, and heat exchanger was considered negligible.

2.2 *Mathematical Modelling of Fuel Cell*

The modelling of a PEMFC stack was done based on the work done [9, 10]. A PEMFC fuel cell reversible cell potential at no load condition that represents the effect of hydrogen and oxygen concentration at specific temperature and partial pressure of hydrogen and oxygen is evaluated as follows [11].

$$E_{\text{nerst}} = E_o - \frac{\Delta S}{nF}(T_{\text{cell}} - T_o) + \frac{RT}{nF}\left[\ln\left(P_{O_2}^{0.5}P_{H2}\right)\right] \tag{1}$$

Where E_o is the cell potential at the standard state ($P_o = 1$atm, $T_o = 298$ K). In most cases the value is taken as 1.229 V. ΔS is the entropy associated with 1 mol H_2 reacting at the standard state. P_{H2} is the partial pressure of the supplied hydrogen and P_{O_2} is the partial pressure of the supplied oxygen at the anode. T_{cell} is the fuel cell operating temperature. On simplification the Nernst equation turns out as follows:

$$E_{\text{nerst}} = E_o - \frac{\Delta S}{nF}(T_{\text{cell}} - T_o) + \frac{RT}{nF}\left[\ln\left(P_{O_2}^{0.5}P_{H2}\right)\right] \tag{2}$$

The partial pressure of oxygen and hydrogen at the cathode and anode can be represented as follows:

$$P_{H2} = 0.5 \times P_{H_2O} \times \left(\left(\exp\left(\frac{-4.192i}{T_{\text{cell}}^{1.334}}\right) \times \left(\frac{P_a}{P_{H_2O}}\right)\right) - 1 \tag{3}$$

$$P_{O_2} = P_{H_2O} \times \left(\left(\exp\left(\frac{-4.192i}{T_{cell}^{1.334}} \right) \times \left(\frac{P_c}{P_{H_2O}} \right) \right) - 1 \right) \tag{4}$$

$$P_{H_2O} = -2.8 + \left(2.95 \times 10^{-2} \times T_c\right) - \left(9.18 \times 10^{-5} \times T_c^{\,2}\right) + \left(1.44 \times 10^{-7} \times T_c^{\,3}\right) \tag{5}$$

$T_c = T_{cell}$-273.15, P_{H_2O} = saturation pressure of water, P_c = pressure at cathode, P_a = pressure at anode i = current density

When the current flows through the fuel cell to generate electricity, the cell potential drops due to the consumption of reactants to form products. Losses in cell voltage occur in three different stages, i.e. activation loss (V_{act}), ohmic loss (V_{ohmic}), and concentration loss (V_{conc}), which significantly depend upon operating temperature and current density. The activation loss occurs due to a part of the overall cell voltage being used to facilitate the slow rate of electrochemical reactions taking place at the anode and cathode. This loss is quite high at a low current density and its effect decreases as the current density increases. The activation losses can be given as:

$$V_{act} = \xi_1 + \xi_2 T_s + \xi_3 T_s . \ln(C_{O_2}) + \xi_4 T_s . \ln(I_{cell}) \tag{6}$$

Here C_{O2} is the oxygen concentration at the cathode which is given by [12],

$$C_{O_2} = \frac{P_{O2}}{5.08 \times 10^6 \times e^{-\frac{498}{T_s}}} \tag{7}$$

ξ is the parametric coefficient that has been calculated by many researchers using different fuel cell dynamic models. The typical values are [12]:

$$\xi_1 = -0.9477, \xi_2 = 3.3 \times 10^{-3}, \xi_3 = 7.5 \times 10^{-5}, \xi_4 = 1.915 \times 10^{-4} \tag{8}$$

The ohmic losses in the fuel cell occur due to the obstruction to the flow of electrons through the outer circuit and the flow of proton through the semi-permeable membrane inside the PEMFC. Therefore, the overall ohmic loss includes both protonic and electronic loss components and can be represented as [12]:

$$V_{ohmic} = I_{cell} \times R_{cell} \tag{9}$$

R_{cell} is the overall resistance offered in the PEMFC which includes the configuration of the fuel cell as well and can be represented as [12]:

$$R_{cell} = 0.001605 - 0.000035T_{cell} + 0.00008i \tag{10}$$

The concentration losses in a fuel cell occur when there is a limitation to the flow of hydrogen and oxygen to the reaction site. This occurs at a higher current density approaching the limiting current density when the rate of reactant consumption is higher than the rate of supply. These losses can be represented as [12]:

$$V_{\text{conc}} = -\frac{RT}{nF}\ln\left(1 - \frac{i}{i_l}\right) \tag{11}$$

Here i_{max} is the maximum current density.

The overall cell potential including the losses is given by:

$$V_{\text{cell}} = E_{\text{nernst}} - V_{\text{ohm}} - V_{\text{con}} - V_{\text{act}} \tag{12}$$

The power generated in the PEMFC stack is calculated as:

$$W_{\text{fc}} = N_{\text{cell}} V_{\text{cell}} I_{\text{cell}} \tag{13}$$

where N_{cell} is the number of stacks and I_{cell} is the current passing through the entire stack. The available power at the fuel cell can be calculated by estimating the energy stored in the hydrogen supplied at the anode site. The theoretical consumption rate of hydrogen in the fuel cell is given by:

$$n_{\text{H2}} = N_{\text{cell}} \frac{I_{\text{cell}}}{2F} \tag{14}$$

The actual hydrogen consumption rate is maintained at a little higher rate for continuous operation of fuel cell while preventing cell degradation. The actual consumption rate is given as:

$$n_{\text{H2}} = \lambda_{\text{H2}} N_{\text{cell}} \frac{I_{\text{cell}}}{2F} \tag{15}$$

Here λ_{H2} is the stoichiometric coefficient of the fuel cell.

The performance parameters of a PEMFC used in this study to solve the above equations are listed in Table 2.

Table 2 Performance parameters of fuel cell

Parameters	Symbol	Value
Number of stacks	N_{cell}	13,000
Current density (operation and limiting), A/cm^2	i and i_l	0.6, 1.5
Faraday's constant, C/mol	F	96,485
Pressure at anode and cathode, bar	P_a, P_c	1.45, 1 respectively
Active surface area, cm^2	A_{cell}	232
Stoichiometric coefficient for hydrogen	λ_{H2}	1.15

2.3 *Performance Evaluation*

The isentropic efficiency of pump and turbine are assumed as 85% and 80% respectively. The thermal efficiency of the coupled system based on the first law of thermodynamics can be calculated as:

$$\text{Thermal efficiency} = \frac{W_{\text{net}}}{Q_{\text{net}}} \times 100 \tag{16}$$

The overall efficiency and the electrical efficiency are given by:

$$\text{PEMFC electrical efficiency} = \frac{W_{fc}}{n_{\text{H2}} \times \text{HHV}} \times 100 \tag{17}$$

$$\text{Overall efficiency} = \frac{W_{\text{fc}} + W_{\text{turbine}} - W_{\text{pump}}}{n_{\text{H2}} \times \text{HHV}} \times 100 \tag{18}$$

3 Results and Discussion

The expansion ratio is the ratio between working pressure of the condenser and the evaporator which is defined as

$$r_p = \frac{P_2}{P_1} \tag{19}$$

Where P_2 the condenser is pressure and P_1 is the evaporator pressure. The expansion ratio is assumed as 0.16.

The variation of pump power consumption in the RORC cycle with the change in expansion ratio for the refrigerant R-245fa at constant evaporator pressure is depicted in Fig. 3.

As the expansion ratio increases, the power consumption by the pump decreases. This is because with the increase in expansion ratio, the condenser pressure increases at constant evaporator pressure. So, the pump has to consume less work to deliver the

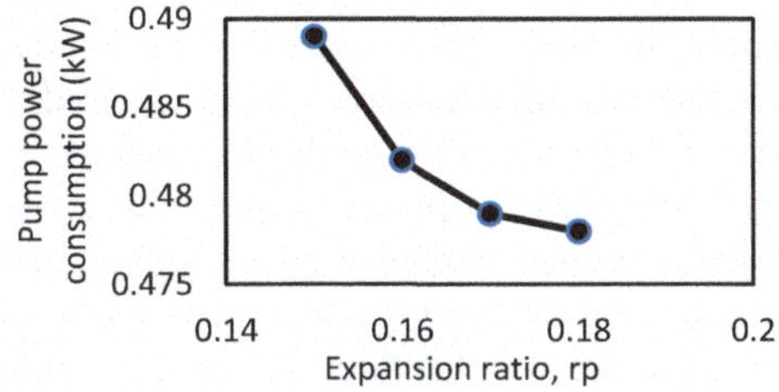

Fig. 3 Pump power consumption in RORC cycle

Fig. 4 Turbine work of RORC with temperature

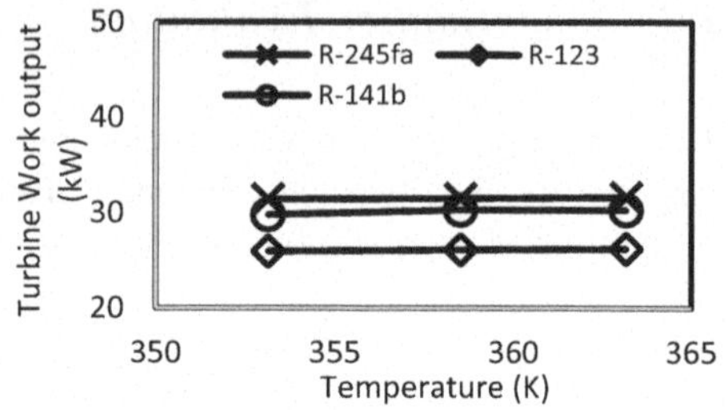

Fig. 5 Thermal efficiency of RORC cycle

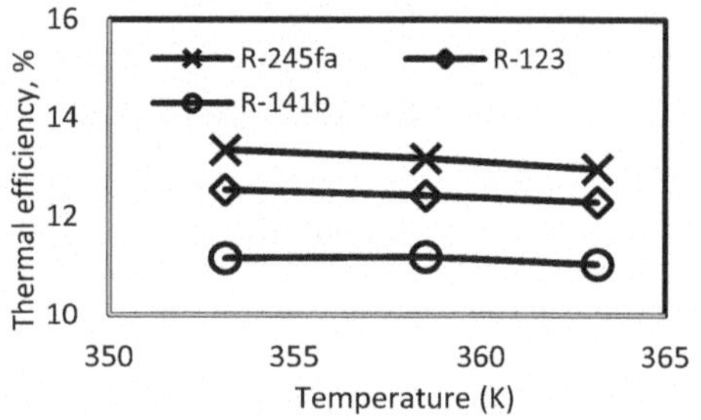

Fig. 6 Overall power output with temperature

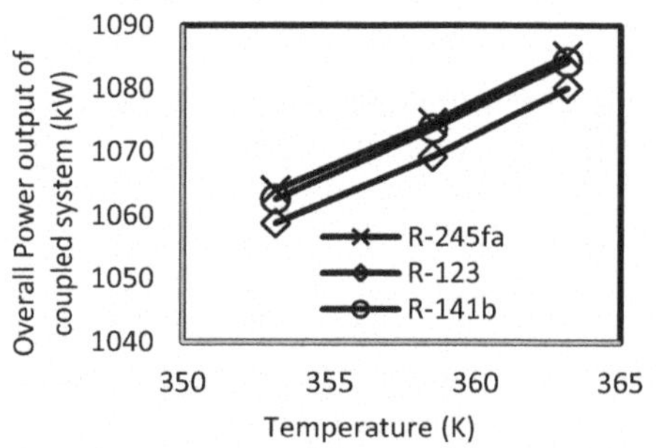

refrigerant to the evaporator pressure. It is seen in Fig. 4 that, with the increase in PEMFC operation temperature, the turbine work marginally increases even though the expansion ratio is maintained as constant at 0.16. The higher operating temperature increases the inlet temperature of the RORC refrigerant entering inside the turbine and hence the maximum working pressure in the turbine increases, which greatly affects the expander work output [12]. The variation in thermal efficiency with temperature for the three different working fluids is depicted in Fig. 5. Figure 6 shows the trend of the overall power output of the coupled system with temperature for the three different working fluids used. It can be observed from the figure that the thermal efficiency of the RORC cycle decreases with the rise in temperature despite the rising trend of power output. This is because as the temperature of the operation increases, the amount of heat supplied in the evaporator is increased at a constant degree of superheating and expansion ratio [12]. The trend of overall efficiency with temperature for the three different working fluids used and three different mass

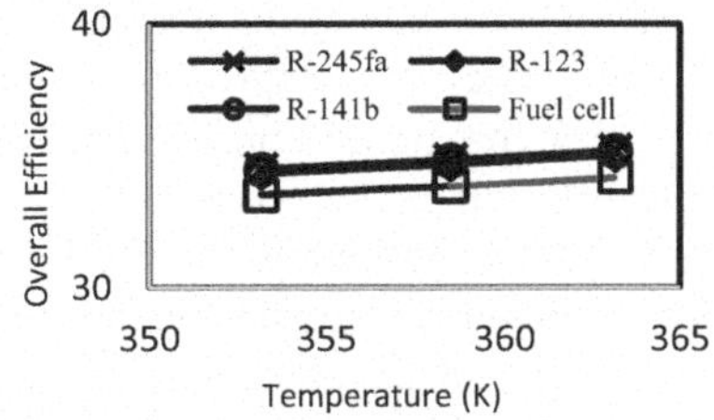

Fig. 7 Overall efficiency with temperature

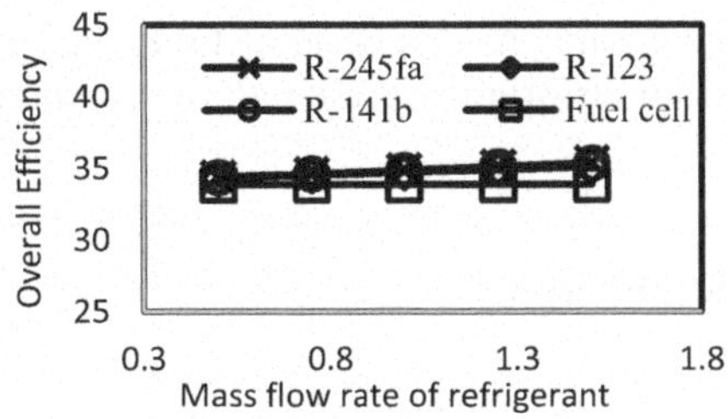

Fig. 8 Overall efficiency with mass flow rate

flowrates in this study is portrayed in Figs. 7 and 8 respectively. The increase in thermal load at higher temperatures results in a reduction in thermal efficiency. But, the power out of the fuel cell keeps on increasing at a greater gradient, and hence there is an improvement in the overall power output (Fig. 7) and overall efficiency (Fig. 8) of the hybrid system temperature.

With the increase in mass flow rate through the RORC cycle, the refrigerant flowing through the RORC system absorbs a greater amount of waste heat and converts them into useful work. As a result, the power output of the overall system increases and consequently improves the overall efficiency of the coupled system than the PEMFC stack working independently. With the increase in the mass flow rate of refrigerant, the improvement in overall efficiency is enhanced. It is to be also noticed that increasing the mass flow rate of refrigerant increases the volumetric flow rate of the refrigerant. The volumetric flow rate is quite high for refrigerants with high liquid density and hence they require larger components size to accommodate the refrigerants [12].

4 Conclusion

The conclusions drawn from this study are the following:

The overall efficiency of the coupled system can be improved at higher temperatures of operation of the fuel cell but maximum improvement may not be achieved at the highest temperature. For the refrigerant R-245fa, the maximum improvement in efficiency of 1.0045% is found at 85 °C whereas for refrigerants R-123 and R-141b, the maximum improvement of 0.8397% and 0.9809% is obtained at 90 °C when the mass flow rate of RORC refrigerant was maintained at a constant value of

1 kg/s. By increasing the mass flow rate of refrigerant to 1.5 kg/s, the improvement in the efficiency of 1.2596% is found at 85 °C whereas for refrigerants R-141b and R-245fa, maximum improvement of 1.4596% and 1.5075% is obtained at 90 °C for a RORC refrigerant mass flow rate of 1.5 kg/s. R-245fa was found to be the best performing refrigerant for the RORC cycle showing the maximum performance improvement at all working conditions. Simulation results indicated that there was no significant improvement in the efficiency of the hybrid system. However, the results may be useful as a basis for future attempts. Experimental investigation is required to assess the performance of the fuel cell-RORC hybrid system is further necessary. At this point of time, the authors are unable to calculate the added cost from the original cycle into the combined one.

References

1. Viswanathan, B., & Aulice Scibioh, M. (2007). *Fuel cells: Principles and applications* (p. 494). Universities Press.
2. Nguyen, H. Q., & Shabani, B. (2020). Proton exchange membrane fuel cells heat recovery opportunities for combined heating/cooling and power applications. *Energy Conversion and Management, 204*, 112328.
3. He, T., Shi, R., Peng, J., Zhuge, W., & Zhang, Y. (2016). Waste heat recovery of a PEMFC system by using organic Rankine cycle. *Energies, 9*(267), 3–15.
4. Yin, Y., Zhao, J., Yin, Y., Zhang, J., & Otoo, O. N. (2020). Thermodynamic modeling and analysis of a novel PEMFC-ORC combined power system. *Energy Conversion and Management, 217*, 112998.
5. Alijanpour Sheshpoli, M., Ajarostaghi, S. S. M., & Delavar, M. A. (2018). Waste heat recovery from a 1180 kW proton exchange membrane fuel cell (PEMFC) system by recuperative organic Rankine cycle (RORC). *Energy, 157*, 353–366.
6. Sai Kaushik, A., Bhogilla, S. S., & Muthukumar, P. (2021). Thermal integration of proton exchange membrane fuel cell with recuperative organic Rankine cycle. *International Journal of Hydrogen Energy, 46*(27), 14748–14756.
7. Babatunde, A. F., & Sunday, O. O. (2018). A review of working fluids for organic Rankine cycle (ORC) applications. *Materials Science and Engineering, 413*, 012019.
8. Mann, R. F., Amphlett, J. C., Hooper, M. A. I., Jensen, H. M., Peppley, B. A., & Roberge, P. R. (2000). Development and application of a generalized steady-state electrochemical model for a PEM fuel cell. *Journal of Power Sources, 86*, 173–180.
9. Khan, M. J., & Iqbal, M. T. (2005). Modelling and analysis of electrochemical, thermal, and reactant flow dynamics for a PEM fuel cell system. *Fuel Cells, 4*, 463–475.
10. Mubin, A. N. A., Bahrom, M. H., Azri, M., & Ibrahim, Z. (2017). Analysis performance of proton exchange membrane fuel cell (PEMFC). *IOP Conference Series; Material Science Engineering, 210*, 012052.
11. Mann, R. F., Amphlett, J. C., Hooper, M. A. I., Jensen, H. M., Peppley, B. A., & Roberge, P. R. (2000). Development and application of a generalized steady-state electrochemical model for a PEM fuel cell. *Journal of Power Sources, 86*(1–2), 173–180.
12. Pratyusha Padhi, S. M. (2021). *Performance analysis of polymer electrolyte membrane fuel cell coupled with recuperative organic Rankine cycle for waste heat recovery*. B. Tech thesis, 1–34.

Interdependency of Passive Design Strategies for Energy-Efficient Building Envelope

Rabee M. Reffat, Murat Cetin, and Adel A. Abdou

1 Introduction

Building activities are energy consumption intensive. Therefore, the building industry and construction sector have become one of the fastest growing energy consuming sectors. Achieving energy efficiency in buildings is quite important since it significantly contributes to the reduction of energy consumption and leads to better utilization of existing energy resources. Building envelope is an important part of the building when it comes to energy efficiency. The primary physical components of the envelope include walls, roof, exterior doors, and windows. Building envelope serves as the outer shell to help maintain the indoor environment and facilitate its climate control. It provides a thermal barrier and filter between the indoor and outdoor environment, and its elements are among the key determinants of a building's energy requirements. Common measures of the effectiveness of a building envelope include physical protection from weather and climate (comfort), indoor air quality, durability, and energy efficiency. Energy minimization relies on the thorough understanding of building thermal behavior, in which the building envelope plays a key role. There are various factors that should be taken into account while addressing the building envelope from energy efficiency perspective. These factors include building shape, orientation, percentage of glazed openings and type characteristics, envelope materials including surface properties and insulation, the building envelope tightness level, adjacent neighboring buildings, shading devices and

R. M. Reffat (✉)
Nottingham University Ningbo China, Ningbo, China
e-mail: rabee.reffat@nottingham.edu.cn

M. Cetin
Kadir Has University, Istanbul, Turkey

A. A. Abdou
KFUPM, Dhahran, Saudi Arabia

X. Wang (ed.), *Future Directions in Energy Engineering*, Green Energy and Technology, https://doi.org/10.1007/978-3-031-62042-3_7

application of intelligent technologies of building envelope. Energy efficiency of a building originates with initial design decisions. Therefore, passive design strategies for energy efficiency need be considered at the early stages of the building design since energy-efficient features are most easily and effectively incorporated at that time. The design process of a building undergoes many phases with different levels of details. This study aims at investigating and identifying a set of passive design strategies that are decided upon by a designer during conceptual designing of buildings. In particular, facades not only offer the building's aesthetic expression, but it can also be advantageously used to control the internal conditions of the building since it represents the building's envelope or "skin" [1].

Related literature to the scope of this paper includes climatic design and rule-based systems. The climatic design approach and its applications in the building sector [2] dates back to the notion of designing in accordance with nature [3, 4] as well as the vernacular concept of passive design principles [5, 6] while the rule-based generic design approaches can be traced back to the ideas of structuralism [7–9] and its ramifications on the linguistic approaches on design process [10, 11]. Manioglu and Yilmaz [12] conducted a study on energy-efficient design strategies in the hot dry area of Turkey derived from a simplified thermal evaluation of traditional and modern houses. Sozer [13] presented the effect of passive solar design techniques for designing the building envelopes to develop and demonstrate high performance. Mwasha et al. [14] investigated the principal of sustainable energy performance indicators for modeling the sustainable performance of the residential building envelope and developed an approach for determining the most appropriate sustainable energy performance indicators. Most recent studies addressed the application of energy-efficient strategies in buildings [15], analysis of design strategy of energy efficient buildings based on databases by using data mining [16], strategic design of wall envelopes for the enhancement of building thermal performance [17], fenestration and window system design methods for high performance buildings [18], and different envelope scenarios towards low carbon high-rise buildings [19].

In this context, the problem area of this study is defined in the intersection between the fields of the formal rules of building envelope design and its energy-efficiency design principles. Therefore, the study is focusing on passive design strategies that help to minimize the energy consumption of buildings in relation to the following key elements: shape and proportions of building envelope, building orientation, envelope materials including surface properties and insulation characteristics, and type and percentage of glazed openings. These strategies are extracted via thorough and comprehensive analysis of the literature and case-analysis of high performance and energy-efficient buildings. The strategies can be utilized by architects at the conceptual stage of architectural design and can be used in developing a computer-aided strategy-based passive design tool for energy-efficient building envelope. The passive design strategies of building envelope are in the form of rules and their interdependencies are analyzed and interpreted.

2 Extraction of Passive Design Strategies for Energy-Efficient Building Envelope

Based on thorough investigation of the related literature on energy-efficient building envelope, passive design rules are extracted in simple postulates as given in Table 1. These rules are clustered into four categories: Shape and Proportion (SP), Orientation (OR), Envelope Materials (EM), and Glazing (GL). There are 51 rules that have been extracted.

3 Interdependency Analysis of Energy-Efficient Building Envelope Rules

The extracted rules are cross-mapped in the form of an interdependency matrix so as to put forward their impact on each other. The rules are arranged in a matrix format whereby their relationships are mapped out as shown in Table 2. The influential level of the relationship of each rule is then determined based on the number of relationships that each rule establishes with the remaining 50 rules. The relationships are estimated based on the embodiment and consequential effect as investigated in the literature that a rule has on each of the other 50 rules.

For example, the rule "SP 01" which describes the use of W/L ratio closer to 1:1 leading to optimum solar insulation is related to, as well as would impact on, the rule "OR 08" which calls for orienting the building plan along East-West axis in order to provide the best use of daylight and subsequently leads to the reduction of energy use due to artificial lighting utilization. Similarly, the rules "OR 03" and "OR 04" addressing the use of rotating technologies of façade shading devices and the need for utilizing orientation-sensitive façade technologies which are interrelated. The rule "OR 03" from the view point of rotating technology is related and affects the rule "EM 17" which is recommending the use of external movable shading devices as part of the envelope materials to provide better energy performance. Another example, the rules "EM 13" and "EM 15" are very closely related. The rule "EM 13" calls for the use of thermal insulation materials in solid exterior walls while the rule "EM 15" specifically recommends the use of Extruded Polystyrene (EPS) as an effective thermal insulation material. Meanwhile, the rule "EM 13" is related to and complements the rule "GL 07" recommending the use of glazing system of thermal conductivity values less than 0.4 to minimize the energy use of air-conditioning in the building.

The obtained results of investigating the interdependency level for each rule in the above matrix are presented in Table 3. For instance, the rule "GL 07" in the category of the Glazing of the Building Envelope (GL) has the most influential level of interdependencies due to having 16 relationships with the other rules out of 50. This is followed by the rules "OR 03" in the category of the Orientation of the Building Envelope (OR), "SP 04" in the category of Shape and Proportion of the

Table 1 Extracted rules for energy-efficient building envelope

	No.	Extracted rules (passive design strategies) for energy-efficient building envelope	References
Shape and Proportion (SP)	SP 01	Width/Length ratio (W/L) closer to 1:1 provides optimum solar insulation	[20]
	SP 02	High internal Volume/external Surface ratio (S/V) provides optimum solar insulation	[21–23]
	SP 03	Low Roof/Wall (W/R) ratio provides reduced thermal stress	[21]
	SP 04	Low built floor area/site area ratio provides better daylighting	[24]
	SP 05	Change in the direction of in/outlet of wind circulation provides increased natural ventilation	[25]
	SP 06	Maximum exposure to summer breeze provides increased natural ventilation	
	SP 07	Minimum exposure to winter winds provides minimized heat loss in winter	[26]
	SP 08	Balanced proportion of heat gains on North and South facades provides optimum energy performance	
Orientation (OR)	OR 01	Design utilization of shading nature for glazing in North façade provides optimum energy performance	[27]
	OR 02	South façade with double façade treatment provides optimum energy performance	
	OR 03	Rotating technologies of façade shading devices provide optimum energy performance	
	OR 04	Orientation-sensitive façade technologies provide optimum energy performance	[28]
	OR 05	Minimum area of openings in East and West facades helps avoiding solar heat gain	
	OR 06	Minimum area of glazing in East and West helps avoiding solar heat gain	
	OR 07	South façade with shading helps reducing solar heat gain	
	OR 08	Orientation of plan along East West axis provides best use of daylighting by reducing lighting energy requirements	
	OR 09	Minimizing façade surface on East and West helps reducing solar heat gain and maximizing daylight transmission	
	OR 10	South façade with stepped profile of self-shading surface helps reducing solar heat gain	

Envelope Materials (EM)	EM01	Ventilated cavity between layers of building envelope provides optimum energy performance	[29]
	EM02	Extraction of heated air between layers of building envelope helps minimizing heat transmission loss	
	EM03	Insulating Concrete Formwork (ICF) wall inherent thermal mass helps reducing heating/cooling load of building envelope	
	EM04	High R-value of wall systems provides optimum energy performance	
	EM05	Locating thermal mass inside an insulated building envelope provides optimum energy performance	[30]
	EM06	Protection of thermal mass from summer sun and cool breezes provides better cooling performance	
	EM07	Air tightness of building envelope helps reduction of heat loss	
	EM08	Walls with single-skin façade contributes to cooling load (hot-humid)	[31]
	EM09	Walls with of double-skin provides better energy performance (in comparison to single-skin facades)	[32]
	EM10	The use of single glazing as inner panel and double glazing as outer panel of double skin reduces the cooling energy requirement	
	EM11	Transparent facades contribute to the increase of cooling loads (in comparison to reflective facades)	[31]
	EM12	Shaded envelope provides better energy performance	[33]
	EM13	Thermal insulation of external walls decreases air-conditioning electric consumption	
	EM14	Ventilation through double-skin facades provides better energy performance	[34]
	EM15	The use of interior EPS (extruded polystyrene) in building envelopes reduces energy consumption	[33]
	EM16	The use of bright louvers in building envelope design reduces energy consumption in summer	
	EM17	External bright and light movable shading devices of climatic facades provide better energy performance	[35]

(continued)

Table 1 (continued)

	No.	Extracted rules (passive design strategies) for energy-efficient building envelope	References
	EM18	Increasing air flow rate in multiple skin building envelope provides better energy performance by reducing thermal load	[36]
	EM19	The use of double reflective glass in single-skin and double absorptive glass in double-skin façade provides better energy performance	[32]
	EM20	Introduction of fan to ventilated shaft of double-skin building envelope provides better energy performance	[34]
	EM21	Higher cavity in ventilated facades helps reducing summer over-heating	[37]
	EM22	Sloping (31-34 degrees) of windows on East and West facades helps reducing direct solar radiation	[38]
	EM23	Lower Window to Wall Ratio (WWR) provides better energy performance	[39]
	EM24	The use of materials with higher thermal conductivity (U-value) provides better energy performance	[40]
	EM25	The use of interior shading devices provides less improvement in energy performance (in comparison to external shading devices)	[41]
Glazing (GL)	GL01	Increasing fenestration area reduces artificial lighting energy consumption	[42]
	GL02	Higher window Visible Transmittance Level (VTL) increases energy savings	[43]
	GL03	Higher window perimeter to floor area ratio increases energy savings	[42, 44, 45]
	GL04	Higher WWR increases solar heat gain	[44, 46]
	GL05	Glazed curtain walls increase energy consumption	[47]
	GL06	Reduction of glazed surfaces (20-30 %) improves thermal comfort	[48]
	GL07	U-value less than 0.4 minimizes air conditioning energy use costs	
	GL08	Lower level of air leakage reduces heat loss/gain (depending on season)	

Table 2 Interdependency matrix of the extracted 51 rules of energy-efficient building envelope

	Shape & Proportions (SP)								Orientation (OR)										Envelope Materials (EM)																									Glazing (GL)										
	SP 01	SP 02	SP 03	SP 04	SP 05	SP 06	SP 07	SP 08	OR 01	OR 02	OR 03	OR 04	OR 05	OR 06	OR 07	OR 08	OR 09	OR 10	EM 01	EM 02	EM 03	EM 04	EM 05	EM 06	EM 07	EM 08	EM 09	EM 10	EM 11	EM 12	EM 13	EM 14	EM 15	EM 16	EM 17	EM 18	EM 19	EM 20	EM 21	EM 22	EM 23	EM 24	EM 25	GL 01	GL 02	GL 03	GL 04	GL 05	GL 06	GL 07	GL 08		TOTAL	
6				X	X			X			X					X	X																																			SP 01	6	Shape & Proportions (SP)
1			X																																																	SP 02	1	
0																																																				SP 03	1	
7											X		X	X			X					X								X		X																				SP 04	8	
2										X			X																																							SP 05	3	
5										X	X		X	X		X																																				SP 06	5	
5											X			X		X																X																		X		SP 07	5	
4											X																				X	X																		X		SP 08	5	
4																																								X					X			X		X		OR 01	4	Oreintation (OR)
0																																																				OR 02	2	
4												X																						X	X												X					OR 03	9	
3																																							X	X										X		OR 04	4	
3														X																										X							X					OR 05	6	
2																																								X							X					OR 06	6	
5																		X																X	X												X	X				OR 07	5	
3																							X	X																									X			OR 08	6	
1																																															X					OR 09	3	
3																														X				X														X				OR 10	4	
1																																																		X		EM01	1	Envelope Materials (EM)
0																																																				EM02	0	
1																																																		X		EM03	1	
1																																																		X		EM04	2	
1																															X																					EM05	2	
1																															X																					EM06	2	
1																																																			X	EM07	1	
4																																					X						X						X	X		EM08	4	
2																																																	X	X		EM09	2	
2																													X																					X		EM10	2	
5																																					X	X			X								X	X		EM11	6	
0																																																				EM12	2	
2																																	X																	X		EM13	5	
0																																																				EM14	3	
1																																																		X		EM15	2	
0																																																				EM16	3	
0																																																				EM17	2	
0																																																				EM18	0	
0																																																				EM19	2	
0																																																				EM20	1	
1																																																		X		EM21	2	
0																																																				EM22	4	
2																																																	X	X		EM23	3	
1																																																		X		EM24	1	
0																																																				EM25	1	
2																																														X	X					GL01	2	Glazing (GL)
0																																																				GL02	1	
0																																																				GL03	1	
1																																																	X			GL04	7	
0																																																				GL05	3	
0																																																				GL06	6	
0																																																				GL07	16	
0																																																				GL08	1	
	0	0	1	1	1	0	0	1	0	2	5	1	3	4	0	3	2	1	0	0	0	1	1	1	0	0	0	0	1	2	3	3	1	3	2	0	2	1	1	4	1	0	1	0	1	1	6	3	6	16	1			

building (SP), and GL 04 in the category of the Glazing of the Building Envelope (GL) because they established 9, 8 and 7 relationships with the other rules respectively. Following this group, the rules "GL 06," "OR 05," "OR 06," "OR 08," "EM 11," and "SP 01" have 6 relationships with other rules. Then, the rules "SP 07," "SP 08," "OR 07," and "EM 13" have 5 relationships with other rules. The rules "OR 01," "OR 04," OR 10," "EM 08," and "EM 22" have established only 4 relationships with other rules. The remaining rules, which are mostly dominated by the rules in the category of Envelope Materials of the Building (EM) have only set 3 or less relationships with other rules. Consequently, some of the rules on "glazing," "orientation," and "shape and proportions" of energy-efficient building envelope stand out with high level of interdependencies with other rules. The rules that are mostly interrelated can be seen to be concentrated in the categories of "orientation" and "shape and proportion" of the building envelope. These interdependency levels could be viewed as the overall influence levels.

Table 3 Top influential levels and interdependency level of extracted rules.

Influential level	Total no. of relationships	Rule #	Category	Interrelationship level of an influential rule to a energy-efficient building envelope category of rules — Interrelationship Level: No of relationships	%	Level
V	6	SP 01	SP	3 out of 6	50%	IV
V	6	SP 01	OR	3 out of 6	50%	IV
III	8	SP 04	OR	4 out of 8	50%	IV
II	9	OR 03	SP	5 out of 9	56%	IV
V	6	OR 05	SP	3 out of 6	50%	IV
V	6	OR 06	SP	3 out of 6	50%	IV
V	6	OR 08	SP	3 out of 6	50%	IV
V	6	**EM 11**	EM	4 out of 6	67%	III
IV	7	**GL 04**	OR	5 out of 7	71%	II
V	6	**GL 06**	EM	5 out of 6	83%	I
I	16	**GL 07**	EM	12 out of 16	75%	II

Level I > 80, Level II 70–80, Level III 60–69, Level IV < 60

On the other hand, the categorical or local influence of each rule is determined based on the number of relationships within its own category. Accordingly, the rules that obtained the highest influential level of interdependencies in relation to its specific category of rules for achieving energy-efficient building envelope are determined. The obtained results as illustrated in Table 3 show that the rule "GL 06" has the highest local interdependency level of 83% interdependency within its "Envelope Materials" (EM) category. This indicates that the rule "GL 06" "Reduction of glazed surfaces (20–30%) improves thermal comfort" is mostly interrelated (highest level) with 5 out of 6 rules in the category of (EM) "Envelope Materials" for achieving energy-efficient building envelope. Also, the rules "GL 04" and "GL 07" are closely related (level II) with their own categories of 71% and 75% interdependency with (OR) "Orientation" and (EM) "Envelope Materials," respectively. While the rule "EM 11" is always related (level III) of 67% interdependency with the other rules in its own category (EM) "Envelope Materials." Furthermore, the rules ("SP 01" and "SP 04") and ("OR 03," "OR 05," "OR 06," and "OR 08") are somewhat related (level IV) of 50% interdependency with rules in the categories of (OR) "Orientation" and (SP) "Shape and Proportion," respectively.

4 Conclusion

The obtained results of interdependency analysis of passive design rules for achieving energy-efficient building envelope presented in this paper identified both the influential and interdependency levels among these rules at both overall and local levels. The rules of "U factor less than 0.4 of glazing," "Rotating technologies of façade shading devices," "Low built floor area / site area ratio," and "window to wall ratio" are the most influential passive design rules in terms of their relationships within the extracted set of 51 rules for achieving energy-efficient building envelope within the context of hot-humid climate zones. Furthermore, the analysis revealed the local level of relationship of some of the passive design rules within their specific categories. For instance, the passive design strategy "Reduction of glazed surfaces (20–30%) improves thermal comfort" has a strong local level of relationship with the passive design strategies in the category of "Envelope Materials."

The extracted set of 51 passive design rules of energy-efficient building envelope in a hot humid climate acquired in this paper are induced based on a thorough investigation of the related literature on strategies for achieving energy-efficient building envelope. The developed interdependency matrix represents the consequential effects, embodiment, and interdependency analysis of passive design strategies for achieving energy-efficient building envelope and provides a systematic reference to support the decision-making process at the early stage of conceptual design towards achieving an energy-efficient building envelope. This matrix can be further refined with inputs from experts in the field as well as results from energy simulation case studies. The extracted rules obtained in this paper will be augmented and refined in forthcoming research with rules to be extracted from energy simulation results using data mining techniques. These refined results will be utilized in developing a passive design strategies computational tool that support achieving energy-efficient building envelope in hot-humid climates.

References

1. Mir, M. (2008). Energy efficient architecture and building systems to address global warming, leadership and management in engineering, engineering strategies for global climate change. *American Society of Civil Engineers ASCE, 8*(3), 113.
2. Watson, D. (1993). *Climatic building design; energy-efficient building principles and practices*. McGraw-Hill.
3. McHarg, I. (1969). *Design with nature*. Wiley.
4. Frommes, B. (1980). *Fundamental knowledge in urban and building climatology*. International Federation of Housing and Planning (IFHP).
5. Tzamtzis, A. (1986). *Passive design; principles for hot and humid climates*. Miami-Dade Community College.
6. Szokolay, S. (1988). *Climatic data and its use in design*. RAIA Education Division.
7. Jakobson, R. (1960). Closing statement: Linguistics and poetics. In T. Sebeok (Ed.), *Style in language*. MIT Press.

8. Foucault, M. (1966). *The order of things: An archaeology of the human sciences*. Gallimard.
9. Levi-Strauss, C. (1968). *Structural anthropology*. Penguin.
10. Stiny, G. (1980). Introduction to shape and shape grammars. *Environment and Planning B; Planning and Design, 7*, 343–351.
11. March, L., & Steadman, P. (1971). *The geometry of the environment; an introduction to spatial organization in design*. RIBA Publications.
12. Manioglu, G., & Yılmaz, Z. (2008). Energy efficient design strategies in the hot dry area of Turkey. *Building and Environment, 43*, 1301–1309.
13. Sozer, H. (2010). Improving energy efficiency through the design of the building envelope. *Building and Environment, 45*, 2581–2593.
14. Mwasha, A., Williams, R., & Iwaro, J. (2011). Modeling the performance of residential building envelope: The role of sustainable energy performance indicators. *Energy and Buildings, 43*(9), 2108–2117.
15. Mousavi, S., Gheibi, M., Wacławek, S., Smith, N., Hajiaghaei-Keshteli, M., & Behzadian, K. (2023). Low-energy residential building optimisation for energy and comfort enhancement in semi-arid climate conditions. *Energy Conversion and Management, 291*(117264), 1–16.
16. He, Y., Chu, Y., Song, Y., Liu, M., Shi, S., & Chen, X. (2022). Analysis of design strategy of energy efficient buildings based on databases by using data mining and statistical metrics approach. *Energy & Buildings, 258*(111811), 1–18.
17. Saboor, S., Chelliah, A., Gorantla, K., Kim, K., Lee, S., Shon, Z., & Brown, R. (2021). Strategic design of wall envelopes for the enhancement of building thermal performance at reduced air-conditioning costs. *Environmental Research, 193*(110577), 1–12.
18. Feng, F., Kunwar, N., Cetin, K., & O'Neill, Z. (2021). A critical review of fenestration/window system design methods for high performance buildings. *Energy & Buildings, 248*(111184), 1–22.
19. Sarogloua, T., Theodosiou, T., Givoni, B., & Meir, I. (2019). Study of different envelope scenarios towards low carbon high-rise buildings in the Mediterranean climate – can DSF be part of the solution. *Renewable and Sustainable Energy Reviews, 113*(109237), 1–12.
20. Ling, C., Ahmad, H., & Ossen, D. (2007). The effect of geometric shape and building orientation on minimizing solar insulation on high-rise buildings in hot humid climate. *Journal of Construction in Developing Countries, 12*(1), 27–38.
21. Behsh, B. (2002). Building form as an option for enhancing the indoor thermal condition. In *Proceedings of the 6th Nordic Building Physics Symposium* (pp. 759–766).
22. Sozer, H., & Clark, R. (2008). Looking for cultural response and sustainability in the design of a high-rise tower in the middle east. In *CTBUH 8th World Congress, Dubai, UAE*.
23. Ratti, C., Raydan, D., & Steemers, K. (2003). Building form and environmental performance: Archetypes, analysis and an arid climate. *Energy and Buildings, 35*, 49–59.
24. Carlo, R., Dana, R., & Koen, S. (2003). Building form and environmental performance: Archetypes, analysis and an arid climate. *Energy and Buildings, 35*, 49–59.
25. Robinette, G. (1983). *Energy efficient site design*. Van Nostrand Reinhold Company.
26. Watson, D., & Labs, K. (1983). *Climatic design: Energy-efficient building principles and practices*. McGraw-Hill/Book Company.
27. Barkkume, A. (2007). Innovative building skins: Double glass wall ventilated Façade. In *Introduction to sustainable design in architecture*. New Jersey Institute of Technology.
28. Sullivan, C. (2006). *Robo buildings: Pursuing the interactive envelope*. Architectural Record.
29. Kosny, J., Christian, J., & Desjarlais, A. (1998). Performance check between whole building thermal performance criteria and exterior wall measured clear wall R-value, thermal bridging, thermal mass and air-tightness. *ASHRAE Transactions, 104*(2), 1379–1389.
30. Ren, W., Lan, M., & Hao, Y. (2006). The study on thermal performance and applicability of energy saving wall materials in hot summer and cold winter zone. In *International conference for enhanced building operations (ICEBO)*.
31. Hamza, N. (2008). Double versus single skin Façade in hot arid areas. *Energy and Buildings, 40*, 240–248.

32. Chan, A., Chow, T., Fong, K., & Lin, Z. (2009). Investigation on energy performance of double skin façade in Hong Kong. *Energy and Buildings, 41*, 1135–1142.
33. Jinghua, Y., Changzhi, Y., & Liwei, T. (2008). Low-energy envelope design of residential building in hot summer and cold winter zone of China. *Energy and Buildings, 40*, 1536–1546.
34. Wong, P., Prasad, D., & Behnia, M. (2008). A new type of double-skin façade configuration for the hot and humid climate. *Energy and Buildings, 40*, 1941–1945.
35. Taylor, P., Duponchel, H., & Bannister, P. (2009). The impact of glazing selection on daylight and energy performance for an office building in Canberra. In *11th international IBPSA conference*.
36. Valentina, S., Fabio, Z., & Marco, P. (2010). Evaluation of a climate façade: Energy efficiency and thermal efficiency. *Energy and Buildings, 42*, 50–62.
37. Carla, B. (2002). A simple model to study ventilated façade energy performance. *Energy and Buildings, 34*, 469–475.
38. Guedi, I. (2003). Energy performance of the self-shading building envelope. *Energy and Buildings, 35*, 327–336.
39. Chou, S., Chua, K., & Ho, J. (2009). A study on the effects of double skin façades on the energy management in buildings. *Energy Conversion and Management, 50*, 2275–2281.
40. Yilmaz, Z. (2007). Evaluation of energy efficient design strategies for different climatic zones: Comparison of thermal performance of buildings in temperate humid and hot dry climate. *Energy and Buildings, 39*, 306–316.
41. Poirazis, H., & Blomsterberg, A. (2005). Energy and thermal analysis of glazed office buildings using a dynamic energy simulation tool. In *Proceeding of the 9th international IBPSA conference* (pp. 945–952).
42. Bodart, M. (2002). Global energy savings in offices buildings by the use of daylighting. *Energy and Buildings, 34*(5), 421–429.
43. Krarti, M., Erickson, P., & Hillman, T. (2005). Simplified method to estimate energy savings of artificial lighting use from daylighting. *Building and Environment, 40*, 747–754.
44. Abdel-Mohimen, M., Hanna, G., & Krarti, M. (2005). Analysis of daylighting benefits for office buildings in Egypt. *Journal of Solar Energy Engineering, 127*(3), 366–370.
45. Ghisi, E., & Tinker, J. (2005). An ideal window area concept for energy efficient integration of daylight and artificial light in buildings. *Building and Environment, 40*, 51–61.
46. Lam, J., & Li, D. (1998). Daylighting and energy performance for air conditioned office buildings. *Energy, 23*, 79–89.
47. Nedhal, A., Al-Tamimi, M., Fairuz, S., & Adel, A. (2009). The effect of orientation and glazed area to the indoor air temperature in unventilated buildings in hot-humid climate. In *3rd international conference on built environment in developing countries (ICBEDC)* (pp. 424–433). Universiti Sains Malaysia, Penang, Malaysia.
48. Melendo, J., & La Roche, P. (2008). Effects of window size in daylighting and energy performance in buildings. In *Proceedings of passive and low energy architecture (PLEA)*.

Harnessing Solar Energy for Efficient Cooling in Hot Weather Regions: A Comparative Experiment on Solar-Assisted Air Conditioning's Energy Efficiency

Sayed Mohamad Soleimani, Mohamad Iyad Al-Khiami, Ahmad Sedaghat, and Mahdi Ashtian Malayer

1 Introduction and Literature Review

The continuous disregard for the environment amongst communities and the pressing need for industrial development have led to a continuous increase of greenhouse gas emissions. This urgency for development has elevated climate change to the forefront of human threats, surpassing poverty, and nuclear weapons [1]. The Middle East and North Africa (MENA) region alone has been found to emit 3.2 billion tons of carbon dioxide which contributes to an average of 8.7% of global warming [1]. The rise of greenhouse gases has intensified the phenomenon of global warming in the world generally and the MENA region. According to Zittis et al. [2], a rise of 0.45°C per decade is expected to continue. Consequently, there has been a significant increase in energy consumption, totaling 24.2% between 2010 to 2020 [3].

When considering Kuwait, a country in the Middle East that exhibits an arid, hot climate, the building sector contributes to most of the energy consumption [4]. According to a report by Kuwait Institute for Scientific Research [5] demand for air-conditioning services contributes to 70% of residential electrical demand. With the integration of renewable energy which currently satisfies only 1% of Kuwait's energy demands [5], solar energy has the potential to significantly

S. M. Soleimani (✉)
The Pennsylvania State University, University Park, PA, USA
e-mail: smsoleimani@psu.edu

M. I. Al-Khiami · A. Sedaghat
Australian University – Kuwait, Mubarak Al-Kabeer, Kuwait

M. A. Malayer
Islamic Azad University, Yazd, Iran

X. Wang (ed.), *Future Directions in Energy Engineering*, Green Energy and Technology, https://doi.org/10.1007/978-3-031-62042-3_8

contribute to the reduction of energy consumption demanded by air conditioners as Kuwait exhibits long durations of sunlight during nearly all seasons.

Solar Assisted Air Conditioners (SAAC) have been perceived as a great solution to reduce energy consumed by air conditioners (AC). Photovoltaic (PV) technology, due to the advancement in its technology, has seen a great reduction in cost and is now considered a significant competitor to fossil fuel-based power generation, not to mention its low environmental impact. Using photovoltaic (PV), that can directly convert solar energy into electricity would decrease the dependency on non-renewable energy, thus reducing the overall energy consumption [6].

Haffaf et al. [6] compared two identical rooms of 20 m^2 in Antalya, Turkey during the month of August. The first room's AC was supported by solar energy while the other was not. The results of the experiment found that energy consumption by the compressor was reduced by 8–28%. On the other hand, a study conducted by Al-Okbi et al. [7] collected solar energy through a vacuum collector and combined it with a conventional air conditioner. The results showed, when compared to conventional AC, a reduction in energy consumption by 52%.

Another study evaluated the performance of a solar hybrid air-conditioner supported by PV cells with R410a refrigerant in comparison to a conventional split unit air conditioner system. The study found substantial improvements in energy consumption. In addition, the study also revealed that solar hybrid air-conditioners are found to be more efficient than conventional split units [8].

According to the authors' best knowledge, the literature review did not explore through an experimental approach (utilizing a control and experimental of identical dimensions), the potential advantages of SAAC in comparison to conventional AC split units in reducing energy consumption specifically in a geographic region like Kuwait.

2 Research Question

The research aims to answer the following research questions (RQ):

RQ1—Does SAAC considerably reduce the total electrical energy consumption when compared with conventional split units in Kuwait?
RQ2—Are SAAC efficient enough to maintain the indoor climate in a hot region?

3 Methodology

This section will start by introducing the geographic location in which the experiment was conducted, followed by the experimental configuration.

3.1 Climate of Kuwait

Kuwait is a country located in the Arabian Peninsula, specifically in the West of Asia. It has a coastline of 500 km. Kuwait latitudes between 28° and 31° North and longitudes between 46° and 49° East. Kuwait summers are extremely hot. For example, during the months of July to August, the average temperatures range between 33° to 47 °C [9] and a record extreme of 54 °C on July 21, 2016 [10].

3.2 Experimental Cabins

Initially, two identical portable cabins were constructed in Kuwait, specifically in Mubarak Al-Kabeer governorate. These cabins are designed to simulate a typical room in Kuwaiti buildings. The internal cabins' dimensions are shown in Fig. 1 and measure 2 m by 2 m, with a ceiling height of 3 m. As for the materials, the walls and the roof are constructed using 76 mm Polyurethane sandwich panels, interior ceiling is covered with a 60 × 60 cm gypsum board grid. The flooring is composed of 18 mm thick plywood with a vinyl finishing.

Each room contains a single leaf door centered, with dimensions of 2 m × 1 m, and a double leaf aluminum window with dimensions of 1 m × 1 m, and a sill height of 1 m. The window glass is weatherproof. Figure 2 shows the cabin's structure at the site.

All wiring necessary for electrical sockets runs through PVC conduits fixed in the wall sandwich panel. Each room has an LED light installed.

3.3 Measuring Devices

In terms of measurement devices, Davis Instrument Airlink sensor was employed to monitor the indoor temperature and humidity levels in real time. The energy consumption on the other hand were measured using Vue Energy Monitor—a device developed by Emporia that provides real-time data on electricity consumption.

3.4 Conventional and Solar-Assisted AC Units

Regarding the air conditioning, a split AC unit with a capacity of 18,000 BTU is installed in each cabin.

The component of the SAAC consists of (1) PV panels, (2) AC unit, (3) batteries, (4) inverter, and (5) power network.

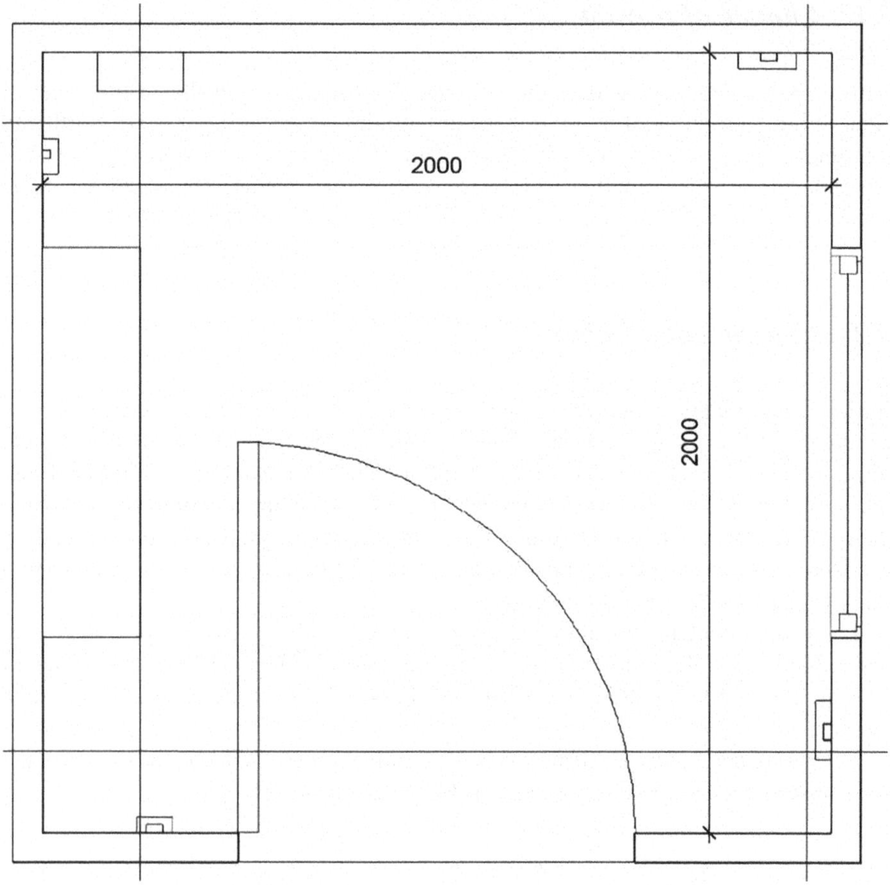

Fig. 1 2D plan view of the constructed cabin

The split unit would get power from each of the three sources by the controller, namely, inverter.

Figure 3 shows the components of the SAAC. Four Monofacial Go Green 350w 24v solar panels were used (total costs 700 USD). An off-grid MPPT inverter charger 3.5KVA 100A 24v was used (total costs 600 USD). Two gel deep cycle batteries 12v 200ah were integrated (total costs 650 USD). Two pole DC miniature circuit breaker 16A, 100A, and 40A were used (total costs 300 USD). All wiring and installation had an average cost of 3250 USD.

Fig. 2 Manufactured cabins in Mubarak Al-Kabeer

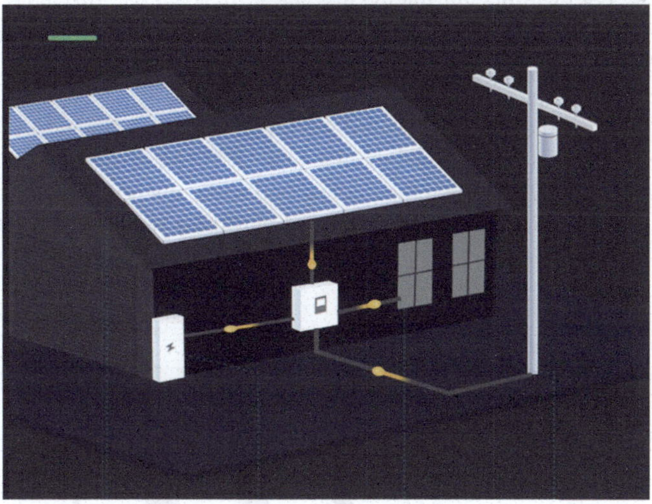

Fig. 3 Sketch showing the four components of SAAC system

4 Results and Discussion

The results of the two AC systems were analyzed by first, pre-setting both the conventional AC unit and the SAAC at 22°C during the months of April and May 2023.

Before proceeding with the data collection, it was crucial to investigate and test the reliability of the sensors and ensure that both cabins have identical temperatures and the AC (in both cabins and prior to the solar energy assistance) have identical energy consumption.

The results of the initial testing have shown identical temperature and energy consumption in both portable cabins [4].

Once the similarity check was confirmed, the SAAC was connected and the data collection was recorded from 26th of April 2023 to 30th April 2023, and then from first of May to 21st of May.

Figures 4, 5, 6, and 7 show the results of the collected data.

From Figs. 4, 5, 6 and 7, during the month of April between the 26th and 30th, the average energy consumption of the left cabin (without SAAC) was around 0.27 kWh. When compared with the right cabin (with SAAC) the average was 0.20 kWh. A difference of 0.07 kWh between the conventional and experimental room was perceived. On average, it can be observed that during those 5 days, an average energy saving of 28% was achieved.

In May 2023, the trend of energy saving between the SAAC and the conventional AC continues. For example, during the first week of May (1st–7th) the energy saved when using SAAC was 31%. Even though the average consumption of the left cabin was 0.19 kWh, that is only 0.01 lower than during the month of May, the saving difference is 3% higher.

When examining the second and third week of May (8th–14th and 15th–21st), the average saving percentage dropped to 24% and 18%, respectively. Both the SAAC and the conventional AC on average consumed more energy than in the previous weeks. The consumption between the second and third week for the left and right

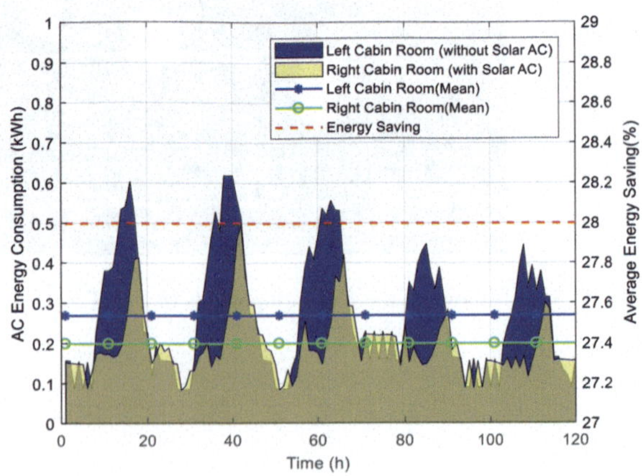

Fig. 4 AC energy consumption against time for the base and SAAC split units during 26th April–30th April

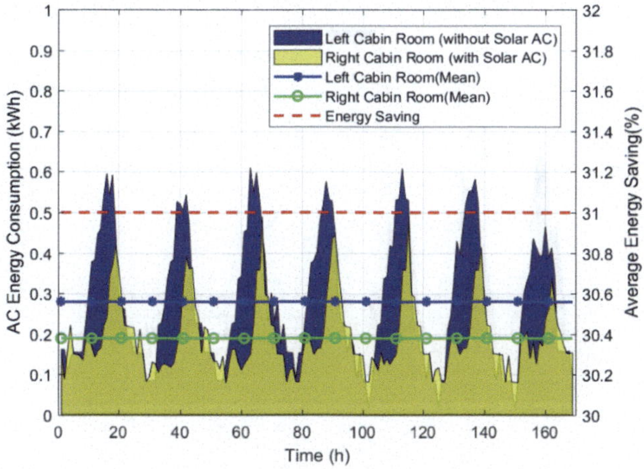

Fig. 5 AC energy consumption against time for the base and SAAC split units during first May–seventh May

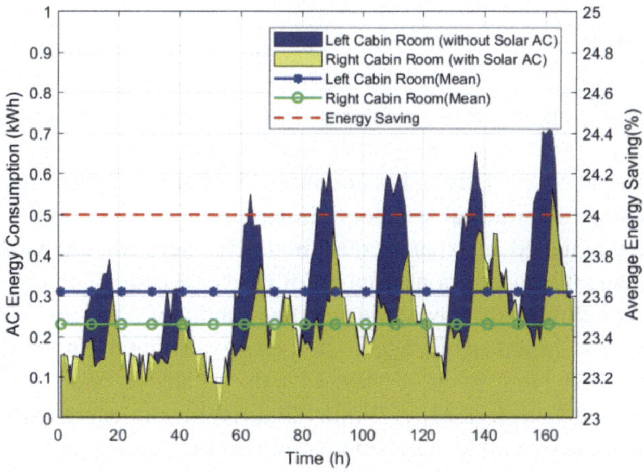

Fig. 6 AC energy consumption against time for the base and SAAC split units during eighth May–14th May

cabins were 0.31 and 0.23 kWh, and 0.29 and 0.24 kWh, respectively. As time passes, the temperatures in Kuwait are continuously increasing, thus maintaining the 22 °C set would require the compressor to work and consume more electrical energy than previous months.

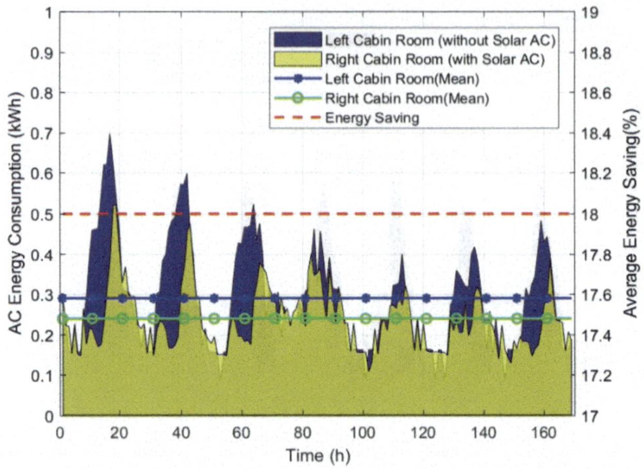

Fig. 7 References AC energy consumption against time for the base and SAAC split units during 15th May–21st May

5 Conclusion

With the continuous increase in greenhouse gas emissions and the concern of global warming that has resulted in a rise of the earth temperature, energy consumptions continue to increase, especially in hot countries like Kuwait, where AC plays a vital role in maintaining a comfortable indoor climate.

The project conducted an experimental approach where two identical cabins were constructed to test the energy consumption of conventional AC in comparison with SAAC in the hot climate of Kuwait during the months of April and May.

The results have shown an average of 28% energy saving during the last 5 days of April, 31% during the first week of May, 24% during the second week of May, and 18% during the third week of May. While the difference in energy saving might be due to the rise of the overall average temperature of the country, SAAC still showed a minimum energy saving of 18% and a maximum of 31%.

The experiment has shown that using solar energy as an electrical source can reduce the total energy consumption of AC while maintaining the same efficiency of conventional AC, thus answering RQ1 and RQ2. It is important to note that further investigation is required during the warmer months of July and August to assess the energy saving that can be achieved.

References

1. Majdi, F., Hosseini, S. A., Karbalaee, A., Kaseri, M., & Marjanian, S. (2022). Future projection of precipitation and temperature changes in the Middle East and North Africa (MENA) region

based on CMIP6. *Theoretical and Applied Climatology, 147*(3–4), 1249–1262. https://doi.org/10.1007/s00704-021-03916-2

2. Zittis, G., et al. (2022). Climate change and weather extremes in the Eastern Mediterranean and Middle East. *Reviews of Geophysics, 60*(3), e2021RG000762. https://doi.org/10.1029/2021rg000762
3. Soleiman Dehkordi, B., & Afrand, M. (2022). Energy-saving owing to using PCM into buildings: Considering of hot and cold climate region. *Sustainable Energy Technologies and Assessments, 52*, 102112. https://doi.org/10.1016/j.seta.2022.102112
4. Sedaghat, A., et al. (2023). Exploring energy-efficient building solutions in hot regions: A study on bio-phase change materials and cool roof coatings. *Journal of Building Engineering, 76*, 107258. https://doi.org/10.1016/j.jobe.2023.107258
5. KISR. *Kuwait energy outlook: Sustaining prosperity through strategic energy management.* KISR. https://www.kisr.edu.kw/media/filer_public/a7/d7/a7d7ecfa-242e-4c5f-a9bc-f971295b0a41/keo_report_english.pdf. Accessed 27 Aug 2023.
6. Haffaf, A., Lakdja, F., Ould Abdeslam, D., & Meziane, R. (2021). Solar energy for air conditioning of an office building in a case study: Techno-economic feasibility assessment. *Renewable Energy Focus, 39*, 148–162. https://doi.org/10.1016/j.ref.2021.09.002
7. Al-Okbi, A., Vankov, Y., & Mohammad Hussain, H. (2021). A hybrid air conditioning system using solar energy to save electrical energy with improving performance. *E3S Web of Conferences, 288*, 01066. https://doi.org/10.1051/e3sconf/202128801066
8. Nwasuka, N. C., Nwankwojike, N., & Nwaiwu, U. (2021). *Performance evaluation of a solar hybrid air-conditioner.* https://doi.org/10.21203/rs.3.rs-503697/v1
9. Sedaghat, A., et al. (2023). Experimental study on cool roof methods for two identical portable cabins in the State of Kuwait. In *The international conference on mechanical, automotive and mechatronics engineering*.
10. Merlone, A., et al. (2019). Temperature extreme records: World meteorological organization metrological and meteorological evaluation of the 54.0°C observations in Mitribah, Kuwait and Turbat, Pakistan in 2016/2017. *International Journal of Climatology, 39*(13), 5154–5169. https://doi.org/10.1002/joc.6132

Optimizing the Urban Rail Transit System for Minimal Passenger Waiting Time and Reduced Traction Energy Consumption

George Yikwanga and Pengfei Sun

1 Introduction

Urban rail transit systems have become the primary mode of transportation for urban residents worldwide, offering safety, punctuality, speed, and convenience [1]. However, the increasing complexity of organizing transportation to meet dynamic passenger flow demands has posed challenges. Adapting and optimizing timetables can improve passenger service and ensure efficient train operations. Aligning the demands of passenger flow and implementing innovative timetable approaches, strategic optimization of train stop times emerges as a practical solution to enhance overall efficiency and safety in urban rail transit systems. Complex timetables are designed to minimize passenger waiting time and prevent crowding to provide public transport services tailored to passenger needs. Optimizing the train operation timetable involves carefully determining departure times, interval operation times, and stop times in response to passenger flow demands and service level standards.

Studies on timetable optimization were initiated by [2] in the early 1970s. Further investigations by Zhang et al. [3] considered influencing factors, establishing both train capacity constraint and no constraint models to minimize travel time, particularly in crowded scenarios. Zhao et al. [4] proposed a multi-objective optimization model aimed at refining subway timetables. Improvements were achieved by prioritizing the minimization of total passenger waiting time, resulting in increased passenger satisfaction. Yunchao et al. [5] proposed an integrated timetable and energy-saving model to minimize the passenger waiting time and the train energy consumption with departure interval as the decision variable. There was a significant improvement in passenger waiting time and energy consumption. In order to reduce the probability of delay, energy consumption and travel time, [6] proposed the

G. Yikwanga (✉) · P. Sun
School of Electrical Engineering, Southwest Jiaotong University, Chengdu, People's Republic of China

X. Wang (ed.), *Future Directions in Energy Engineering*, Green Energy and Technology, https://doi.org/10.1007/978-3-031-62042-3_9

synchronization model of the timetable and stop plan. According to Sun et al. [7], increased accuracy in estimating passenger waiting times is crucial for establishing viable timetables in varying capacity situations. Adjustments to arrival time, running time, stop time, and departure time for specific trains on a given line can reduce overall passenger waiting times and energy consumption. The study introduced a timetable optimization model aimed at minimizing total passenger wait times and energy consumption within urban rail transit systems. Xie et al. [8] proposed the robust optimization approach that minimizes the passenger waiting time and maximizes the utilization of the regenerative energy. Sun et al. [9] presented the optimization model of the metro trains to conserve energy and reduce passenger waiting time by considering a multi-particle train operation model and employing deep reinforcement learning and NSGA-II algorithm resulting in a 9.16% reduction in passenger waiting time and a 15.88% reduction in energy consumption.

The optimization of train operation systems for tractive energy conservation is a difficult task that involves analyzing the relationship between running time and tractive energy consumption [10]. Essentially, any alterations made to train timetables, based on running times, certainly influence not only the timetable itself but also the energy consumed by trains. Introducing alternative running times can generate a range of potential energy-efficient timetable options. A method proposed by Vandanjon et al. [11] introduces an approach capable of generating alternative running times resulting in minimal tractive energy consumption. This approach to timetable optimization is based on the concept of an energy-efficient speed profile, a concept redefined by Howlett [12] to determine the most precise optimal solution. According to Su et al. [10], traction energy consumption is directly linked to running time. Reducing running times leads to an increase in energy consumption, while longer running times reduce tractive energy consumption. The overall operating time for a complete journey includes the minimum required running time, additional operating time (often called time supplements), and stop times [13]. Optimizing stop times can impact overall traction energy consumption. Any excess time gained from this optimization can be allocated to running time, thereby increasing energy efficiency. Goverde et al. [14] formulated a model aimed at distributing running time supplements across successive trips, resulting in a significant reduction in total energy consumption. Previous studies emphasized the importance of the optimal distribution of running time supplements to minimize energy consumption. This paper aims to contribute to this field by focusing on optimizing stop times to conserve energy while considering passenger waiting times.

2 Establishment of the Model

The selected study line is a one-directional subway line that focuses only on the direction with the highest passenger demand. There are n stations on the line, and the train starts at the first station and goes through each station.

The direction under research observes the operation of n trains, each leaving at intervals defined at the first station n, within the designated study time [0, T]. The optimization goal is to minimize the passenger waiting time and traction energy consumption, figuring out how long each train will stop at each station even when it does not stop. Some assumptions are made, the running times and stop times are given and zero stop time is implemented. We examine trains moving on a level gradient without factoring in regenerative energy. This scenario involves a single speed limit between stations.

2.1 *Modelling Train Movement*

The motion of a train moving from one station to the next can be described using Newton's second law of motion. If the train's length is relatively short and factors such as gradient and curvature have negligible effects, we can treat the train as a single point mass. In the case of a subway train, which is typically not very long, we can express the equations of motion as follows:

$$v\frac{dv}{dx} = \frac{F(v) - B(v) - W_0 - W_l}{M(1+\gamma)} \tag{1}$$

$$\frac{dt}{dx} = \frac{1}{v} \tag{2}$$

where:

v is the speed of the train.
x is the position of the train.
t is the running time of the train.
(v) is the traction force. Which can be expressed as

$$F(v) = k_t F(v)_{\mathrm{m}} \tag{3}$$

where $F(v)_{\mathrm{m}}$ is the maximum tractive force, and is the tractive force adjustment coefficient, which is between [0,1] depending on the operation of the train.

$B(v)$ represents the braking force as follows:

$$B(v) = k_t B(v)_{\mathrm{m}} \tag{4}$$

where $B(v)_{\mathrm{m}}$ is the maximum braking force, and is the braking force adjustment coefficient, which is between [0,1] depending on the operation of the train.

W_0 is the basic running resistance of the train.
W_l is the line resistance.

M is the mass of the train.
γ is the rotary mass coefficient.

The train's basic running resistance can be formulated based on the Davis equation as follows:

$$W_0(v) = a + bv + cv^2 \tag{5}$$

where a, b, c are the coefficient which are non-negative constants and depend on the train characteristics and the line resistance is represented as follows:

$$W_l(x) = Mg\sin(\theta(x)) \tag{6}$$

where $\theta(x)$ is the track slope at position x.

The tractive energy consumption can be formulated as follows:

$$E(T) = \sum_{i=1}^{N}\sum_{s=1}^{K}\int_{S_i}^{S_{i+1}} \frac{F(v)}{\eta} dx. \tag{7}$$

Subjected to the following constraints:

$$0 \leq F(v) \leq F(v)_{\max}$$
$$0 \leq B(v) \leq B(v)_{\max}$$
$$0 \leq v(x) \leq V(x)_{\max}$$
$$v(S_i) = 0$$
$$v(S_{i+1}) = 0$$
$$(T)_{i,\min} \leq t(S_{i+1}) - t(S_i) \leq (T)_{i,\max}$$

the relationship between traction energy consumption and running time is shown in Fig. 1, which indicates that the highest energy consumption occurs during the shortest running time.

2.2 Timetable Model

Given a known Origin-Destination (OD) passenger waiting for train i to reach station s, while when train i leaves at s station at time, t_{di} the waiting time of the passenger is

$$W_{si} = t_{di} - t_a \tag{8}$$

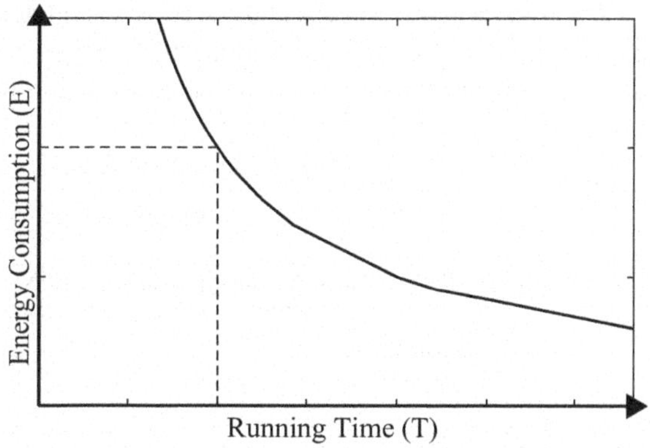

Fig. 1 Energy consumption vs running time. (Adapted from Su et al. [10])

where W_{si} is the waiting time of the passenger, t_{di} is the departing time of the train and the t_a, is the time the passenger arrived at the station.

Considering the total number of passengers from station s to station k at time t, the total waiting time for this group of OD passengers can be expressed as follows:

$$TW_{si} = \sum W_{si} \times N_{si} \tag{9}$$

where TW_{si} is the total waiting time for the group of passengers (OD) and the N_{si} is the number of the waiting passengers at the station. To ensure the punctual departure of the first and last trains, the model introduces a departure time constraint for the first and last train. This constraint represents the departure time constraint

$$t_{sd} = 0 \text{ and } t_{fd} = T \tag{10}$$

where $t_{sd}$is the time for first train departure and t_{fd} is time for last train departure. To ensure the safety of train operation, the model also includes maximum and minimum running interval constraints

$$\text{MinInterval} \leq t_{di,s} - t_{di-1,s} \leq \text{MaxInterval}, \quad i = 2, 3, \ldots, N \tag{11}$$

The arrival time constraints for trains arriving at the same station is expressed as follows:

$$t_{ai,s} - t_{di-1,s} \geq \text{ArrivalTime}, i = 2, 3, \ldots, N \tag{12}$$

where t_{di-1} is the departure time from the previous station and t_{ai} is the actual arrival time at the current station. Stop time constraints is

$$\text{MinStopTime} \leq \text{StopTime} \leq \text{MaxStopTime}, i = 2, 3, \ldots, N \tag{13}$$

The time of train i arrives at the station s is related to the departure time of train i at the previous station and the running time between the stations, while the departure time of train i at the current s station is related to the stop time,

$$t_{ai,s} = t_{di-1,s-1} + \text{RunTime}, i = 2, 3, \ldots, N \tag{14}$$

$$t_{di} = t_{ai,s} + \text{StopTime}, i = 2, 3, \ldots, N \tag{15}$$

$$\text{TotalTripTime(T)} = \sum \text{MinRunTime} + \sum \text{RunTimeSupplements} + \sum \text{StopTime} \tag{16}$$

3 Objective Function

The first objective (f_1) is optimization of the total waiting time for the passengers to board the train, the second objective (f_2) is to minimize the traction energy consumption.

$$f_1 = \min \sum W_{si} \times N_{si} \tag{17}$$

$$f2 = \text{Min } E(T) \tag{18}$$

The two objectives transformed into a single objective function as follows:

$$F = \text{Min}(w1.f1 + w2.f2) \tag{19}$$

4 Simulation Analysis

Using genetic algorithm to optimize train stop times in 0–60-min intervals within a total 38-min running time. Analysis covers 7 trains, 7 sections forming 8 stations during 17:00–18:00 rush hour. Factors considered: train capacity (1000), OD passengers, entry time, weight coefficient (1). Train specifics: distance = [7926, 4234, 7283, 4723, 3242, 3252, 4077] (*m*), running time = [540, 300, 420, 300, 240, 240, 240] (*s*), speed limit: 18 m/s. Safety guidelines maintain a 60-s interval between trains. Other specifics: max traction force (*F*) = 200 kN, max braking force (*B*) = 159.6 kN, running resistance (*R*) = 40 kN, train mass = 216(*t*), motor efficiency = 0.85. Genetic algorithm parameters: population (25), iterations (200), crossover/mutation rate (0.8), selection rate (0.8). The fixed stopping time is 30 s for all stations.

Table 1 Optimized stopping time(s)

Station	Train 1	Train 2	Train 3	Train 5	Train 6	Train 7
S1	25	15	15	15	15	15
S2	15	15	15	15	0	30
S3	25	50	15	15	25	15
S4	15	15	25	15	15	0
S5	15	15	25	25	15	15
S6	15	15	25	15	25	40
S7	35	15	35	50	50	35
S8	25	50	40	35	15	50

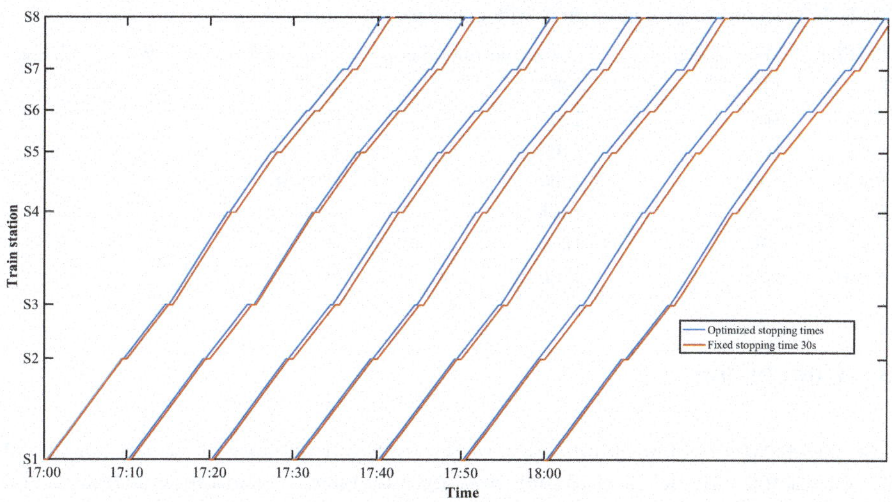

Fig. 2 Timetable with optimized stop time and fixed stop time

Table 1 indicates the optimized stop time results. Further, Fig. 2 presents the comparison of the fixed and optimized time stop time.

Total Waiting Time (f_1) When examining total waiting time with fixed stop times of 30 s it led to passengers waiting for 148,623 min. After optimizing the timetable with adjusted stop times, the waiting time decreased to 142,770 min, representing 3.94% improvement.

Energy Consumption ($f2$) In the case where fixed stop times (30 s) are employed, Table 2 shows the highest traction energy consumption, which occurs due to minimal running time. The table also displays energy consumption when running time supplements are utilized to optimize energy consumption. By implementing only optimized stop times, traction energy consumption can be decreased by 5% in comparison to the fixed stop time, as shown in Table 3.

Table 2 Energy consumption with fixed stop time

Section	Min. run time(s)	Total stop time(s)	Max energy (kWh)	Energy with time supplements (kWh)
S1-S2	3083.01	210	3625.86	2805.95
S2-S3	1647.24	210	1936.76	1404.13
S3-S4	2832.97	210	3331.72	3205.79
S4-S5	1837.43	210	2160.48	1851.57
S5-S6	1261.47	210	1482.88	990.57
S6-S7	1265.39	210	1487.43	999.67
S7-S8	1586.2	210	1864.87	1754.62

Table 3 Traction energy consumption with optimized stop time

Section	Total stop time(s)	Optimized run time(s)	Optimized energy (kWh)	Change
S1-S2	115	3895	2694.17	4%
S2-S3	105	2205	1280.58	9%
S3-S4	160	3100	3159.07	1%
S4-S5	100	2200	1801.56	3%
S5-S6	125	1805	890.54	10%
S6-S7	150	1830	912.06	9%
S7-S8	260	1940	1736.97	1%

5 Conclusion

The paper explores the optimization of the train timetable of the urban rail transit systems using a model, considering passenger demand, scheduling constraints, and behavioral aspects. Through a specialized genetic algorithm, the research had two objectives, minimizing the passenger waiting and reducing the traction energy consumption. The study found an improvement of 3.94% in total passenger waiting times, signifying a substantial increase in efficiency. Moreover, there was a 5% decrease in traction energy consumption. These results highlight considerable progress in improving the passenger experience while at the same reducing the energy consumption within urban rail transit systems. However, this paper's research focuses only on a one-way line as an example. Looking ahead, future research could involve optimizing the operation scheme of a two-way line in real-world scenarios. Additionally, exploring the utilization of regenerative energy could be an important aspect to consider in this area of future study.

References

1. Yang, X., Li, X., Ning, B., et al. (2016). A survey on energy-efficient train operation for urban rail transit. *IEEE Transactions on Intelligent Transportation Systems, 17*(1), 2–13.

2. Amit, I., & Goldfarb, D. (1971). The timetable problem for railways. *Developments in Operations Research, 2*(01), 379–387.
3. Zhang, T., Li, D., & Qiao, Y. (2018). Comprehensive optimization of urban rail transit timetable by minimizing total travel times under time-dependent passenger demand and congested conditions. *Applied Mathematical Modelling, 58*, 421–446.
4. Zhao, L., Li, K., & Su, S. (2013). A multi-objective timetable optimization model for subway systems. In *Proceeding of the International Conference on Electrical and Information Technologies for Rail Transportation* (pp. 557–565).
5. Yunchao, Q., Huan, W., Jianjun, W., Xin, Y., Haodong, Y., & Li, Z. (2020). Robust optimization of train timetable and energy efficiency in urban rail transit: A two-stage approach. *Computers & Industrial Engineering, 146*, 106594.
6. Xie, J., Zhang, J., Sun, K. Y., Ni, S. Q., & Chen, D. J. (2021). Passenger and energy-saving oriented train timetable and stop plan synchronization optimization model. *Transportation Research Part D: Transport and Environment., 98*, 1361–9209.
7. Sun, H., Wu, J., Ma, H., Yang, X., & Gao, Z. (2019). A Bi-objective timetable optimization model for urban rail transit based on the time-dependent passenger volume. *IEEE Transactions on Intelligent Transportation Systems, 20*(2), 604–615.
8. Hadi, S., Hamed, A., Saeed, M., & Hani, G. (2023). Reducing passenger waiting time and energy consumption for metro timetabling: A robust optimization approach. *Energy Reports, 10*, 4051–4062.
9. Sun, Z., He, D., He, Y., Shan, S., & Zhou, J. (2023). A Bi-objective optimization model of metro trains considering energy conservation and passenger waiting time. *Journal of Cleaner Production, 437*, 140427.
10. Su, S., Li, X., Tang, T., et al. (2013). A subway train timetable optimization approach based on energy-efficient operation strategy. *IEEE Transactions on Intelligent Transportation Systems, 14*(2), 883–893.
11. Vandanjon, P., Lejeune, A., Chevrier, R., et al. (2016). Towards eco-aware timetabling: Evolutionary approach and cascading initialization strategy for the bi-objective optimization of train running times. *IET Intelligent Transport Systems, 10*(7), 483–494(.
12. Howlett, P. G. (1990). An optimal strategy for the control of a train. *Journal of the Australian Mathematical Society Series B, 3*, 454–471.
13. Jin, B., Sun, P., Wang, Q., et al. (2020). Two-step method to reduce metro transit energy consumption by optimizing speed profile and timetable. *IET Intelligent Transport Systems, 14*(9), 1097–1107.
14. Goverde, R. M., Bešinović, N., Binder, A., Cacchiani, V., Quaglietta, E., Roberti, R., & Toth, P. (2016). A three-level framework for performance-based railway timetabling. *Transportation Research Part C: Emerging Technologies, 67*, 62–83.

Potentiality of Agricultural Residues from Pruning Fruit Trees for Energy Source in Northern Thailand

Suwimon Wicharuck, Atipoang Nuntaphan, and Chatchawan Chaichana

1 Introduction

Agricultural biomass has been considered a suitable resource for energy production in any country where the economy is based on agriculture [1]. Biomass from agricultural byproducts can be harvested from different sources, for example, crop residues after harvesting or pruning wood from permanent fruit plantations [2]. Almost 60% (28 million ha) of Thailand's area is occupied by agricultural fields such as paddy rice, maize, and fruit trees [3]. Common biomass such as rick husk, corn cob, and sawdust are currently marketed via different trading routes. On the other hand, biomass from the pruning of permanent fruit plantations is not common because it contains high moisture content and requires intense management (such as wood processing and transportation).

Fruit and tree cultivation are ranked the second highest among total agricultural areas in Thailand (approximately 12 million ha). Longan and mango are commonly planted around the country. Longan is mostly cultivated in the northern part of the country, with a total planted area of 194,000 ha in 2021 [4]. Also, mango is an important economic crop in Thailand, and it occupies a total planted area in the north of around 51,965 ha in 2021 [5]. Regular pruning of longan and mango trees is necessary annually in order to reduce insect and disease problems, facilitate orchard

S. Wicharuck
Office of Research Administration, Chiang Mai University, Chiang Mai, Thailand

Mae Moh Training Center, Electricity Generating Authority of Thailand, Lampang, Thailand

A. Nuntaphan
Energy Technology for Environment Research Center, Faculty of Engineering, Chiang Mai University, Chiang Mai, Thailand

C. Chaichana (✉)
Mae Moh Training Center, Electricity Generating Authority of Thailand, Lampang, Thailand
e-mail: c.chaichana@eng.cmu.ac.th

X. Wang (ed.), *Future Directions in Energy Engineering*, Green Energy and Technology, https://doi.org/10.1007/978-3-031-62042-3_10

management, and improve fruit quality. These practices have led to the accumulation of substantial residual amounts. Based on calculations, the biomass production from longan and mango pruning operations could exceed 531 × 10^6 t/year, equivalent to 226,256 ktoe [4–6]. This represents a substantial amount that needs to be disposed of annually.

Currently, the wood residues from pruning are used as firewood for cooking or charcoal production. However, in many cases, the wood will be burnt after cutting in order to remove the residues from the orchard. The combustion of these materials can lead to severe air pollution, including the emission of PM 2.5 particles and haze. This contributes to the escalation of global warming issues and the release of greenhouse gases. Finding an effective way to utilize the wood residue from tree pruning can transform the problem of disposal into a profitable opportunity. Furthermore, the estimation of available residues obtained from pruning management could be used as a source of energy production, supporting supplementary revenue for farmers, and leading to a sustainable ecological system [7–9]. Therefore, the objective of this research was to investigate the availability and fuel properties of pruning wood from longan and mango trees as a source of alternative energy production for thermal biomass power plants in Lampang province, Northern Thailand.

2 Methodology

2.1 Sample Collection

The study was conducted from 2019 to 2020 in Lampang Province, Northern Thailand, around Mae Moh, Mueang Lampang, Ko Kha and Mae Ta districts (Fig. 1). Two main types of fruit varieties were selected for the study as longan (Litchi chinensis Sonn.) and mango (Mangifera L.), according to land use data in 2019 (LDD, 2019). The total covered area of longan and mango was 3104 and 705 ha, respectively. Field surveys and farmer interviews were conducted to gather basic information about fruit tree characteristics, crop calendar, and pruning practices. Sampling sizes were 196 ha of longan and 44 ha of mango, according to the Taro Yamane equation [10].

The selected orchards for pruning wood collection were done at different times due to the harvesting periods of each fruit type. The number of investigations was 25 trees per fruit variety. After the pruning operation, the pruned wood samples were taken and weighed directly by a 20 kg weighting balance. Then, the fresh pruning wood was shredded by a woodchipper. The woodchip size varied from 0.5 × 0.5 to 2.5 × 2.5 cm in width and length, approximately. Shredded woodchip samples of 5 kg from each fruit tree were collected after shredding directly for proximate and ultimate analysis. The analyzed properties were moisture content, ash content, volatile matter, heating values, fixed carbon, carbon, hydrogen, nitrogen, total sulfur, and oxygen.

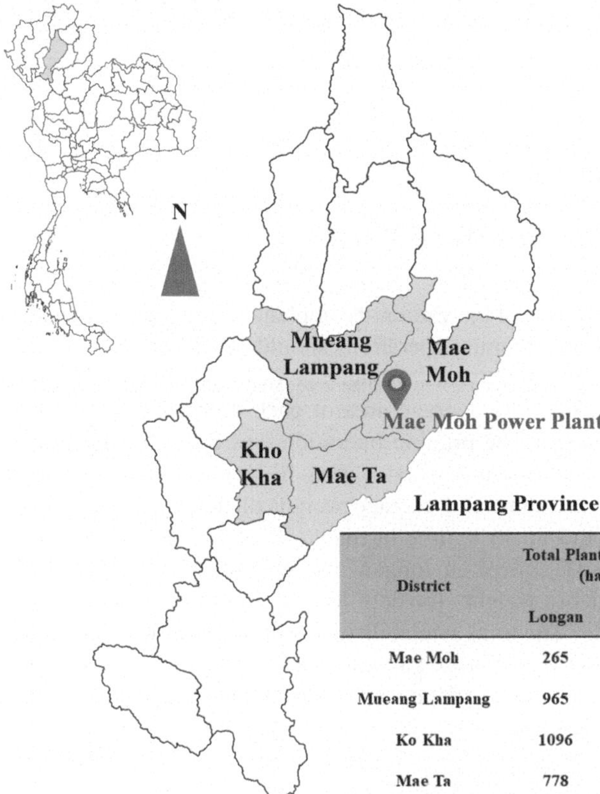

District	Total Planted areas (ha)		Sampling Areas (ha)	
	Longan	Mango	Longan	Mango
Mae Moh	265	124	17	8
Mueang Lampang	965	318	61	20
Ko Kha	1096	86	69	5
Mae Ta	778	177	49	11
Total (ha)	3104	705	196	44

Fig. 1 Location of study areas in four districts of Lampang Province, Northern Thailand

2.2 Power Plant Size Estimation

The total availability of fresh pruned wood (t/year) from the fruit trees in each year was calculated as: pruning wood (kg/tree) × numbers of trees in 1 ha × total planted areas of each fruit variety (ha). The number of fruit trees was 156 per ha, with a spacing between trees of 8 × 8 m (recommended by the DOAE [11]). Also, the installed capacity of thermal biomass power plants (IC, MW) was estimated according to Eq. 1.

$$\begin{aligned} \mathrm{IC} = {} & [(\text{Total dried woodchip} \times \mathrm{LHV}) \div 3.6] \div \text{operation hour} \\ & \times \text{plant efficiency} \end{aligned} \tag{1}$$

Where, operation hours in a year were determined as 300 working days per year and 24 h per day (300 day/year × 24 h/day = 7200 h/year). The efficiency of biomass power plants was 27%, according to the study of Phoumin et al. [12].

3 Results and Discussion

3.1 *Availability of Pruned Wood Weight*

Figures 2 and 3 show the results of the variability of planted fruit trees, crop calendar and pruned wood weight. The pruning operations should be carried out once a year. The age of the fruit trees was 5–40 year (average age of longan trees =18 year and mango trees = 15 year), with the average area of each orchard being 1.07 ha in longan and 0.79 ha in mango. The pruning operation was done directly after fruit harvesting. The diameter of pruning wood varied from one orchard to another, ranging between 10 and 150 mm. The average weight of fresh pruning wood was 36.48 kg/tree in longan and 26.43 kg/tree in mango, with values of pruned wood weight of 22–70 to 17–41 kg/tree in longan and mango, respectively (Fig. 2). Furthermore, dried woodchip weights for both longan and mango were 22.81 and 13.44 kg/tree, respectively. The total availability of pruned wood was 20,571 t/year for both fruit trees (17,664 t/year in longan and 2907 t/year in mango). The total pruned values of each district were 2019 t/year in Mae Moh, 6802 t/year in Mueang Lampang, 6592 t/year in Ko Kha and 5158 t/year in Mae Tha (Fig. 3).

3.2 *Pruning Wood Properties*

Table 1 shows the ultimate and proximate properties of woodchip from longan and mango at both conditions, as received and on a dry basis. For the proximate properties, the results showed that the values of moisture content were 41.8% in longan and 52.3% in mango at as received and 6.9% in longan and 6.2% in mango at

Fig. 2 Average values of fruit tree age (year), fresh pruned wood weight (kg/tree) and dried woodchip (kg/tree)

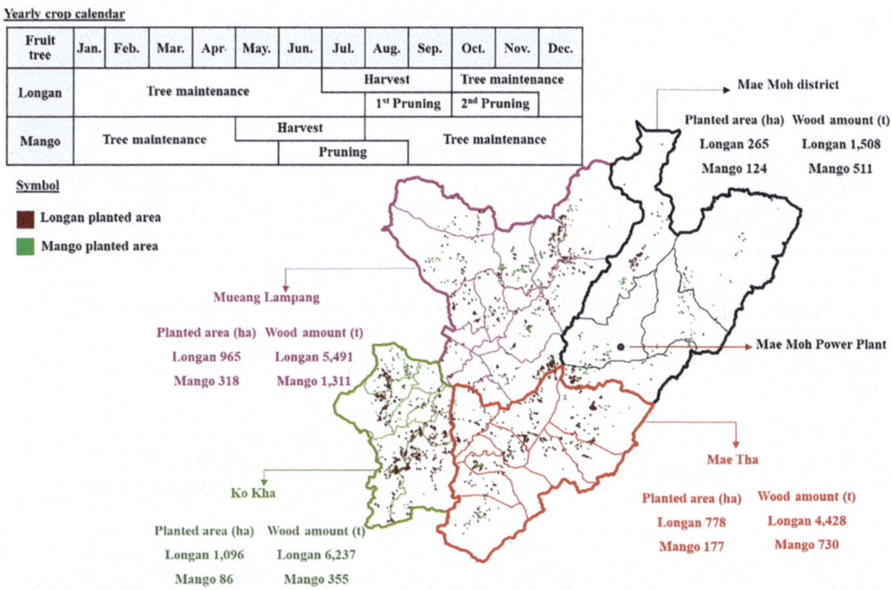

Fig. 3 The yearly crop calendar, total planted areas of longan and mango trees, and the amount of total fresh pruned wood weight of each fruit variety in the study area

dry basis. The ash content on a dry basis ranged from 2.5% in longan to 3.8% in mango. Volatile matter values on a dry basis were 80.1 and 80.6% in longan and mango, respectively. Moreover, the heating values of longan (at dry basis: HHV = 19.3 MJ/kg and LHV = 18.0 MJ/kg) were slightly higher than those of mango trees (at dry basis: HHV = 18.9 MJ/kg and LHV = 17.8 MJ/kg).

According to the ultimate analysis results (Table 1), organic elemental components such as carbon, fixed carbon, nitrogen, oxygen, hydrogen, and sulfur were reported on a dry basis. The values of C (48.8%), H (6.0%), and N (0.3%) in longan woodchip samples were higher than those values in woodchip samples of mango ($C = 47.8\%$, $H = 5.9\%$, $N = < 0.00\%$). The sulfur content in woodchip samples was 0.09% for both fruit varieties. On the other hand, a slightly lower value of oxygen content in longan (42.4%) was observed in comparison to the value in mango (42.5%).

3.3 Potential of Biomass Power Plant Capacity

The results showed that the calculated installed capacity of a thermal biomass power plant (IC) at a plant efficiency of 27% was 2.4 MW, with 2.1 MW of longan and 0.3 MW of mango. This IC was based on 100% dried woodchip from both fruit varieties at a working hour of 7200 h/year. Pruned wood was a seasonal availability,

Table 1 Proximate and ultimate analysis of longan and mango woodchips

Parameters	Values				Method
	Longan		Mango		
	a.r.	d.b.	a.r.	d.b.	
Proximate analysis					
Moisture content (%wt)	41.8	6.9	52.3	6.2	ASTM D3302/3302 M-17
Ash content (%wt)	1.4	2.5	1.8	3.8	ASTM D7582-15
Volatile matter (%wt)	46.6	80.1	38.5	80.6	
Fixed carbon (%wt)	10.1	17.4	7.4	15.6	
Gross calorific value (HHV, MJ/kg)	11.3	19.3	9.0	18.9	ASTM D5865-13
Net calorific value (LHV, MJ/kg)	9.5	18.0	7.2	17.8	
Ultimate analysis					
C (%wt)	–	48.8	–	47.8	ASTM D7582-15
H (%wt)	–	6.0	–	5.9	ASTM D5373-16
N (%wt)	–	0.3	–	<0.00	
S (%wt)	–	0.1	–	0.1	ASTM D4239-18$^{\varepsilon 1}$
O (%wt)	–	42.4	–	42.5	100%- (C + H + N + S)

a.r. and d.b.—as received and dry basis, respectively
C, H, N, S and O were carbon, hydrogen, nitrogen, sulfur, and oxygen
HHV and LHV—high heating values and lower heating values

and the wood might not be available in the same amounts as in the study years. Therefore, a minimum of a 2 MW thermal biomass power plant can be expected in the study area, and the calculation of required thermal energy was 7407 kW_{th}, with a biomass consumption of 1.49 t/h (Table 2).

The availability of pruned wood weight was different from one place to another, depending on biomass types. However, the collection of pruned wood from the agricultural fields has many limitations due to the characteristics and properties of the trees, for example, field-based residues, variations in wood size and high moisture content after cutting. Management of field-based residues from the orchards to the biomass power plant is one important factor for fuel prices, as pointed out by Biocca et al. [9]. The center for wood collections should be set up to gather the wood before shipping it to the power plant. Also, pre-treatment of the pruning wood before transportation, such as size reduction and drying, is needed in order to decrease the costs of storage and logistics and improve the quality of the woodchip. This was also mentioned by Eisenbies [13] that density, moisture content and shape of the wood are the most important physical properties for conversion processes and logistics, as the bulk density of woodchip was about 2–5 times lower than that of wood log, with density values of around 100–250 kg/m^3 for woodchip and 550 kg/m^3 for wood log [6, 14, 15].

Table 2 The installed capacity of a thermal biomass power plant (MW) according to the pruning wood residue (dried woodchip) in the study areas

Detail	Unit	Dried woodchip	
		Longan	Mango
1. Planted areas of each fruit variety[a]	ha	3104[a]	705[a]
2. Pruning wood weight	kg/tree	22.81[b]	13.44[b]
3. Number of trees[c] = 156 tree/ha	kg/ha	3559	2096
4. Annual pruned wood weight	t/year	11,047	1478
5. LHV	MJ/kg	18.0	17.8
1 MJ = 0.001 GJ 1 kWh = 3.6 MJ	GJ/year	198,843	263,050
1 kWh = 0.001 MWh	MWh/year	55,234	7307
6. Operation hour = 7200 h/year (300 day; 24 h/day(	kW_{th}	7671	1015
7. Plant efficiency[d] = 27%	kW_e	2071	274
8. Installed power generation capacity	**MW**	**2.1**	**0.3**
		2.1+0.3=2.4	
9. Required thermal energy	**kW_{th}**	**7,407**	
10. Biomass consumption	**t/h**	**1.49**	

[a]Land use data from Land Development Department, Thailand [3]
[b]Calculated from $M_w = (w{-}m_d) \div m$: when, M_w, w and m_d represented moisture content (%), fresh weight and dry weight, respectively
[c]Spacing between trees = 8 m [11]
[d]Cited from Phoumin et al. [12]

4 Conclusion and Recommendations

Total pruned wood weight from longan and mango fruit trees in four districts of the study area showed significant amounts that indicated a substantial potentiality as alternative energy sources for thermal biomass power plants. However, the pruning wood from longan and mango was identified as a field-based byproduct that was scattered throughout the areas and yearly seasonal availability. According to their availability and properties, pruned wood is difficult to collect from the fields due to its features, such as its size and moisture content. Therefore, a center of wood collection must be implemented in the areas to gather the wood, and pruning wood must be treated as transforming the wood into woodchips and drying before being used. These activities can help to reduce the cost of transportation, and the price of fuel per unit tends to decrease as well. Further investigations on the topic of economic evaluation, optimum technology of thermal biomass power plants, and environmental constraints are needed. Another study on different types of biomass fuel resources should be taken into account as an alternative source of green energy as well. It is recommended that the study should be carried out further in order to develop a general model for estimating the amount of pruning wood available and the consequent biomass power plant size in other areas that may contain different biomass species.

Acknowledgments This research was funded by The Electricity Generating Authority of Thailand (EGAT). Supports from the Office of Research Administration, Energy Technology for Environment Research Center, Faculty of Engineering and Faculty of Agriculture, Chiang Mai University, Chiang Mai, Thailand are acknowledged.

References

1. Cardoen, D., Joshi, P., Diels, L., Sarma, P. M., & Pant, D. (2015). Agriculture biomass in India: Part 1. *Estimation and characterization. Resources, Conservation, Recycling, 102*, 39–48.
2. Dyjakon, A., & García-Galindo, D. (2019). Implementing agricultural pruning to energy in Europe: Technical, economic and implementation potentials. *Energies (Basel), 12*, 1513.
3. LDD. (2018). *Land use map of Lampang Province 2018*. Land Development Department, Ministry of Agriculture and Cooperatives.
4. OAE. (2022). *Cultivated area of longan in Thailand*. Office of Agricultural Economics, Ministry of Agriculture and Cooperatives.
5. DOAE. (2022). *Statistics of mango production in Thailand*. Department of Agricultural Extension, Ministry of Agriculture and Cooperatives.
6. Nagle, M., Habasimbia, K., Leisa, H., Mahayotheeb, B., Janjaic, S., Haewsungcharern, M., & Müller, J. (2007). Availability and potential of local biomass resources as fuel for drying of tropical fruits in northern Thailand. In *Conference on International Agricultural Research for Development* (pp. 1–5).
7. Drigo, R.. (2006). Wood fuel Integrated Supply/Demand Overview Mapping (WISDOM) methodology. Spatial Wood Fuel Production and Consumption Analysis of Selected African Countries. FAO-Forestry Department–Energy, Consultancy Report.
8. Soria-González, J. A., Tauro, R., Alvarado-Flores, J. J., Berrueta-Soriano, V. M., & Rutiaga-Quiñones, J. G. (2022). Avocado tree pruning pellets (persea americana mill.) for energy purposes. Characterization and quality evaluation. *Energies, 15*, 7514.
9. Biocca, M., Gallo, P., & Sperandio, G. (2022). Potential availability of wood biomass from urban trees: Implications for the sustainable management of maintenance yards. *Sustainability, 14*, 11226.
10. Yamane, T. (1973). *Statistics: An introductory analysis*. Harper and Row.
11. DOAE. (2017). *Manual of schedules and instructions for data collection at sub-district level*. Department of Agricultural Extension, Ministry of Agriculture and Cooperatives.
12. Phoumin, H., Kimura, S., Wongsapai, W., & Achawangku, Y. (2019). *Biomass supply chain for power generation in southern part of Thailand* (EIA Research Project Report No. 09).
13. Eisenbies, M. H., Volk, T. A., Therasme, O., & Hallen, K. (2019). Three bulk density measurement methods provide different results for commercial scale harvests of willow biomass chips. *Biomass and Bioenergy, 124*, 64–73.

14. Jensen, P. D., Hartmann, H., Böhm, T., Temmerman, M., Rabier, F., & Morsing, M. (2006). Moisture content determination in solid biofuels by dielectric and NIR re-flection methods. *Biomass and Bioenergy, 30*(11), 935–943.
15. Koppejan, J., & Van Loo, S. (2012). *The handbook of biomass combustion and co-firing*. Routledge.

Microwave-Assisted Pretreatment of Lignocellulosic Biomass with a Low Transition Temperature Mixture: Choline Chloride/Glycerol

Sanphawat Phromphithak, Tossapon Katongtung, and Nakorn Tippayawong

1 Introduction

The development for sustainable and alternative energy sources has become important due to the increasing global energy demands and environmental concerns. There are agreements of global cooperation aimed at reducing the traditional fossil fuels [1]. These efforts encompass both biological and chemical technologies, which have emerged as promising pathways toward the development of new bio-based industries. These technologies hold the potential to reduce greenhouse gas emissions and promote sustainable and environmentally friendly alternatives. The production of alternative energy from lignocellulosic biomass has gained global attention. It focuses on synthesizing green chemicals through the conversion of lignocellulosic biomass.

Lignocellulosic biomass typically comprises of three major components: cellulose (40–80%), hemicellulose (15–30%), and lignin (10–25%). Cellulose is a high molecular weight, semi-crystalline polysaccharide with β-1,4-glycosidic bonds. Hemicellulose is similar to the cellulose, but has lower molecular weight, an amorphous polysaccharide, and various functional groups. Lignin consists of aromatic rings in a long polymeric chain, primarily connected by β-O-4 ether bonds [2]. The covalent bonding between hemicellulose and lignin forms the lignin-carbohydrates complex. This complex contributes to the recalcitrance of biomass leading to resistant in the biomass.

S. Phromphithak
Graduate Program in Energy Engineering, Faculty of Engineering, Chiang Mai University, Chiang Mai, Thailand

T. Katongtung · N. Tippayawong (✉)
Department of Mechanical Engineering, Faculty of Engineering, Chiang Mai University, Chiang Mai, Thailand

X. Wang (ed.), *Future Directions in Energy Engineering*, Green Energy and Technology, https://doi.org/10.1007/978-3-031-62042-3_11

Biomass pretreatment purpose of opening up the plant cell walls breaks down the bonding network between lignin-carbohydrates complex. This process led to in the removal of lignin from the biomass while preserving carbohydrates for subsequent conversion into simple sugars [3]. The obtained carbohydrates were known as cellulose-rich material (CRM). The CRM can be utilized through biomass valorization methods, including gasification [4], hydrothermal processes [5], and pyrolysis [6]. These alternative technologies enable the conversion of CRM into valuable biofuels, bio-chemicals, and other high-value products. Biomass pretreatment plays a role in improving the accessibility and digestibility of lignocellulosic biomass enhancing the overall effectiveness of product conversion processes.

Ionic liquids (ILs) have been in the attention as effective solvents for lignocellulosic biomass due to their unique properties, including non-flammability, low vapor pressure, and chemical stability. However, their practical application on a large scale raises concerns about production costs [7]. To address these issues, low transition temperature mixtures (LTTMs) have emerged as an alternative to ILs. It can be prepared by combining two chemicals that form a liquid through hydrogen bond interactions. The LTTMs offer several advantages such as ease of preparation and cost-effectiveness [8]. Choline derivatives are commonly used as hydrogen bond acceptors in LTTMs due to their favorable properties, low cost, and availability. Glycerol is produced as a by-product in biodiesel production. The increasing global demand for biodiesel has led to a surplus of glycerol, resulting in a decrease in its price [9]. As a result, glycerol has become an economically viable and sustainable chemical for incorporation into LTTMs. This approach shows a reduction in the reliance on expensive ILs in biomass pretreatment processes.

Microwave-assisted liquefaction has gained recognition in the processing of lignocellulosic biomass. In comparison to conventional liquefaction methods, microwave-irradiation heating offers several advantages, including time efficiency, reduced chemical and energy consumption, and economic feasibility [10]. Furthermore, the products obtained through this technique have been reported as a high conversion in many cases studies. Zehui J. et al. [11] utilized a mixture of glycerol and ethylene glycol for the microwave-assisted extraction of wood residues. They conducted the reaction at a temperature of 160 °C for 20–40 min. The results showed significant decomposition of lignin and hemicellulose in the wood, along with an increase in cellulose content. In a similar study by Meng L. et al. [12], microwave-assisted heating was applied to bamboo residues. Increasing the temperature from 75 to 120 °C resulted in a reduction of lignin content. The lignin removal led to an increase in cellulose content. The microwave-irradiation heating demonstrated high performance in liquefaction, primarily aimed at the removal of lignin and hemicellulose while preserving the cellulose crystallinity.

In this study, the corncob (CC) pretreated by microwave-irradiation heating using the LTTM was investigated. The CC was also explored to represent the agricultural residues. The LTTM was used choline chloride combine with glycerol. The microwave-irradiation was varied temperature (90 and 150 °C) and residence time (5 and 10 min). The cellulose, hemicellulose, and lignin contents in the CRMs were determined in accordance with the NREL/TP-510-42623 standard. The study involved understanding and discussions on investigation factors, and a comparison between microwave-irradiation and conventional heating methods.

2 Materials and Methods

2.1 Biomass Sample and Reagents

Choline chloride (≥98%), glycerol (≥99%), acetone, and deionized water (DI water) were acquired from Sigma-Aldrich. Low transition temperature mixtures (LTTM) were created by mixing ChCl with glycerol, stirring at 100 °C for 1 h to obtain the LTTM-Cl. Corncobs (CC) were gathered from a cultivated area in Nan, Thailand. These biomass samples were milled and sieved to achieve particle sizes of approximately 1–5 mm, followed by drying at 105 °C for 24 h in an oven.

2.2 Microwave Heated Biomass Pretreatment with LTTM

The microwave-irradiation heating setup used 10% biomass with LTTM-Cl into a round-bottom with connecting to condenser. It was placed to the microwave-reactor as shown in Fig. 1. To control the temperature in the microwave-irradiation process, a setup involving two Erlenmeyer flasks, one inside and one outside the microwave, was employed. The LTTM-Cl blank was initially placed inside the inner Erlenmeyer flask. During the reaction, the inner flask, containing the LTTM-Cl blank, was moved outside the microwave and monitored using a thermocouple of type K to measure the temperature. Once the measurement was taken, the LTTM-Cl blank was returned to the microwave and the reaction continued until completion. After the reaction, the CRMs and crude LTTMs in all experiment were separated by filtration, while the CRMs still retains some LTTMs. The CRM was washed with a mixture of

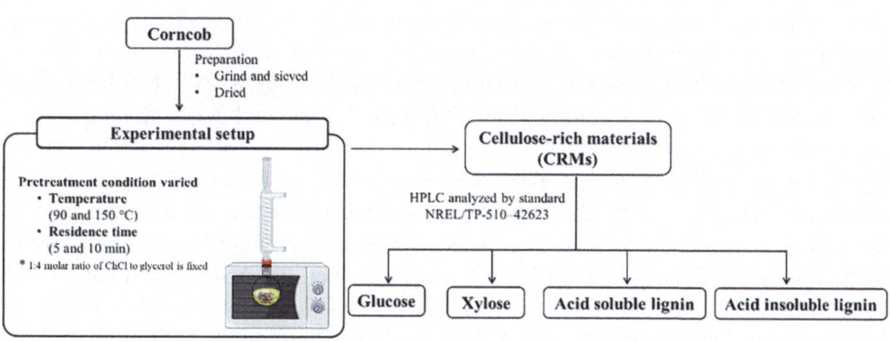

Fig. 1 Schematic diagram for the experiment of corncob pretreatment with LTTM-Cl

deionized water and acetone (50 vol.% concentration) in a centrifuge spinning at 10000 rpm until it was got rid of remaining LTTMs. The obtained CRMs was dried in an oven at 105 °C for 24 h.

2.3 Lignocellulosic Biomass Analysis

The analysis of CC and CRM samples included determining the composition of cellulose, hemicellulose, and lignin. In this context, glucose was used to represent cellulose, xylose represented hemicellulose, and lignin by a combination of acid-soluble lignin (ASL) and acid-insoluble lignin (AIL). The total lignin content was calculated as the sum of ASL and AIL. These analytical procedures were carried out following the NREL/TP-510-42623 standard. Solid recovery (SR), cellulose enrichment factor (CEF), hemicellulose enrichment factor (HEF) in CRM, as well as the ASL removal AIL removal, and lignin removal were calculated as follows:

$$\text{Solid recovery (\%)} = \frac{\text{Weight of CRM (g)}}{\text{Weight of corncob (g)}} \times 100 \tag{1}$$

$$\text{CEF/HEF} = \frac{\text{Cellulose/Hemicellulose contained in CRM (\%w/w)}}{\text{Cellulose/Hemicellulose in corncob (\%w/w)}} \tag{2}$$

$$\text{ASL/AIL/Lignin removal (\%)} = 1 - \frac{\text{ASL/AIL/Lignin contained in CRM (\%w/w)} \times \text{SR}}{\text{ASL/AIL/Lignin contained in corncob (\%w/w)}} \tag{3}$$

3 Results and Discussion

Figures 2 and 3 illustrated the effect of microwave-irradiation heating on corncob pretreatment at varying temperatures and residence times. The ASL removal indicated that temperature had an obvious effect on ASL extraction, with 150 °C resulting in over 90% removal compared to about 50% at 90 °C. Residence time had a minor effect on ASL removal. The AIL removal showed that both temperature and residence time influenced AIL extraction. At a low temperature of 90 °C for 5 and 10 min, ASL removal increased from 40% to almost 60%. At a high temperature of 150 °C for 5 min, the ASL removal is higher than the low temperature of 90 °C for 10 min. It seemed that higher temperatures might break down certain interaction bonds in lignocellulosic biomass. Moreover, the ASL removal at 150 °C for 10 min was higher than for 5 min, rising from around 60% to 80% leading to the result that residence time affected ASL extraction. Consequently, the higher ASL and AIL removal resulted in an increase in the CEF. The CEF increasing was similar

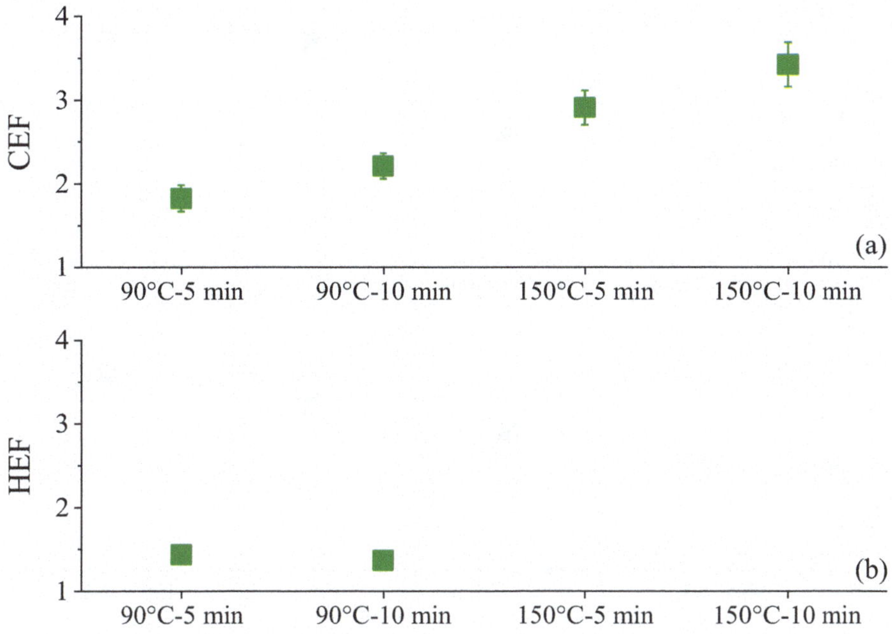

Fig. 2 Effect of microwave-irradiation heating on corncob pretreatment with LTTM-Cl at 90 and 150°C for 5 and 10 min (**a**) CEF and (**b**) HEF

trend to AIL removal, which reflected the effect of temperature and residence time on the cellulose content in CRM. The microwave-irradiation heating method demonstrated that the combination of LTTM-CL with microwave irradiation did not disrupt the cellulose structure. However, hemicellulose content exhibited different behavior. HEF revealed that at a low temperature of 90 °C for 5 and 10 min, there was an increase in hemicellulose content in CRM, and residence time had an insignificant effect on hemicellulose extraction. At a high temperature of 150 °C for 5 and 10 min, the HEF indicated non-detectable hemicellulose content. It implied that the hemicellulose had been removed from CRM. Hemicelluloses were largely preserved within the temperature range of 60–120 °C due to the preservation of side chains within the hemicellulose structure [13]. In a similar study conducted by Pengyun et al. [14], they developed a pretreatment method for lignocellulose fractionation in Pinus bungeana Zucc using deep eutectic solvents (choline chloride and lactic acid). The pretreatment process involved microwave irradiation at 800 W, with the temperature increasing from room temperature to 120 °C over 2 min and then being held at 120 °C for 8 min. The results demonstrated the remarkable effectiveness of 42.81% lignin removal, and HEF of 0.39. Additionally, the method largely preserved cellulose as indicated by CEF of 1.38.

The key parameters compared between conventional heating and microwave-irradiation heating were summarized in Table 1. From our previous study [15], it was determined that the optimal conditions of conventional heating involved a

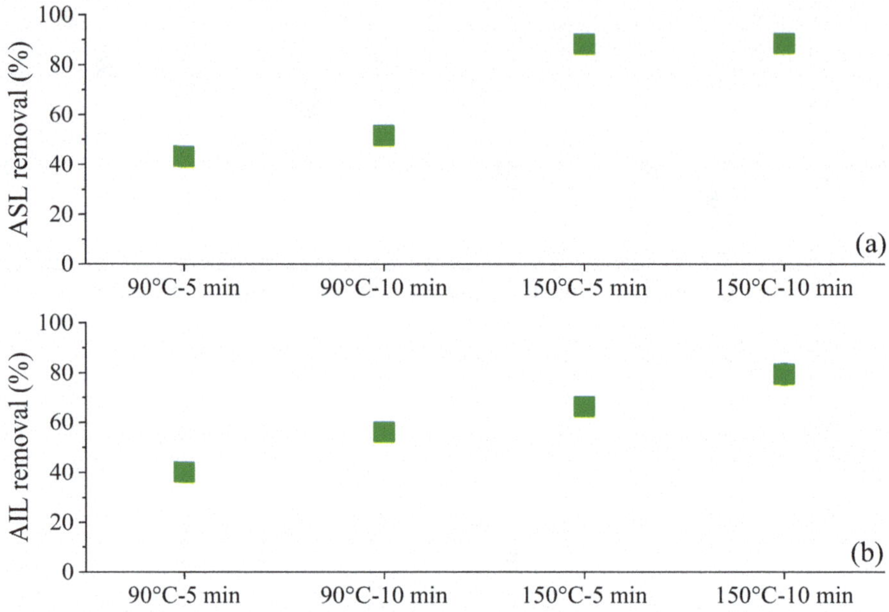

Fig. 3 Effect of microwave-irradiation heating on corncob pretreatment with LTTM-Cl at 90 and 150°C for 5 and 10 min (**a**) ASL removal and (**b**) AIL removal

Table 1 Comparison between conventional heating and microwave-irradiation heating

Method	LTTM	CEF	HEF	Lignin removal (%)	Ref.
Corncob, conventional heating 150°C for 12 h	ChCl to glycerol molar ratio of 1:4	1.53	0.79	27.42	[15]
Corncob, microwave-irradiation heating 90°C for 10 min	ChCl to glycerol molar ratio of 1:4	2.21	1.37	55.40	This work
Corncob, microwave-irradiation heating 150°C for 10 min	ChCl to glycerol molar ratio of 1:4	3.43	Non-detect	81.65	This work
Garlic skin, microwave-irradiation heating 80°C for 20 min	ChCl to glycerol molar ratio of 1:2	1.63	1.00	71.36	[17]

temperature of 150 °C, a residence time of 12 h, and a ChCl to glycerol molar ratio of 1:4. The microwave-irradiation heating outperforms conventional heating in the fractionation of lignocellulose. It achieves efficient fractionation with a shorter residence time and higher CEF [16]. The choice of HEF allows for the preservation or non-preservation with controlling the temperature reaction. Moreover, the microwave-irradiation heating resulted in high removal of both ASL and AIL indicating effective lignin extraction. The microwave-irradiation heating with ChCl to glycerol was also documented by Qinghua J. et al. [17]. They performed pretreatment on garlic skin using a 1:2 molar ratio of ChCl to glycerol at 80 °C for

20 min. The result showed an increase in CEF, along with largely lignin removal. The mild condition at 80 °C can preserve hemicellulose, with a HEF value of 1.00. Their findings are similar to our work. Despite its effectiveness in lignocellulose fractionation, it is worth noting that microwave-irradiation heating is currently employed at the laboratory scale. Scaling up for industrial applications would require further investigation [18]. The analysis of lignocellulosic properties following the NREL/TP-510-42623 standard may not provide a comprehensive assessment of the material. Further analyses, such as X-ray Diffractometer (XRD) to investigate cellulose crystallinity or Thermogravimetric Analysis (TGA) to explore the thermal stability profile, are essential for a deeper understanding of CRM properties. The understanding material properties assists in approaching their suitability for various applications.

4 Conclusion

Biomass pretreatment using LTTMs has proven effective in fractionating lignocellulose. It made them more accessible for subsequent conversion processes. The microwave-irradiation heating highlighted the importance of temperature affecting the extraction of ASL and AIL. Higher temperatures resulted in greater ASL and AIL removal leading to an increased CEF. The hemicellulose preservation was more achievable at lower temperatures, while higher temperatures facilitated its extraction. The optimal composition of CRM can be achieved by modifying pretreatment parameters to suit specific application requirements. The obtained CRMs hold great potential for various downstream applications.

Acknowledgments This work was funded by the National Research Council of Thailand (NRCT), Friend Energy Partnership Ltd., and Fundamental Fund 2566, Chiang Mai University. The first author (SP) wishes to acknowledge a PhD scholarship from the NRCT Research & Researchers for Industry program.

References

1. Tadesse, H., & Luque, R. (2011). Advances on biomass pretreatment using ionic liquids: An overview. *Energy & Environmental Science, 4*(10), 3913–3929.
2. Faraco, V., & Hadar, Y. (2011). The potential of lignocellulosic ethanol production in the Mediterranean Basin. *Renewable & Sustainable Energy Reviews, 15*(1), 252–266.
3. Chen, Y., & Mu, T. (2019). Application of deep eutectic solvents in biomass pretreatment and conversion. *Green Energy & Environment, 4*(2), 95–115.
4. Sasujit, K., Homdoung, N., & Tippayawong, N. (2022). Non-thermal plasma removal of naphthalene as tar model compound from biomass gasification. *Energy Reports, 8*, 97–103.
5. Saengsuriwong, R., Onsree, T., Phromphithak, S., & Tippayawong, N. (2021). Biocrude oil production via hydrothermal liquefaction of food waste in a simplified high-throughput reactor. *Bioresource Technology, 341*, 125750.

6. Khuenkaeo, N., Phromphithak, S., Onsree, T., Naqvi, S. R., & Tippayawong, N. (2021). Production and characterization of bio-oils from fast pyrolysis of tobacco processing wastes in an ablative reactor under vacuum. *PLoS One, 16*(7), e0254485.
7. Halder, P., Kundu, S., Patel, S., Setiawan, A., Atkin, R., Parthasarthy, R., et al. (2019). Progress on the pre-treatment of lignocellulosic biomass employing ionic liquids. *Renewable & Sustainable Energy Reviews, 105*, 268–292.
8. Francisco, M., González, A. S. B., García de Dios, S. L., Weggemans, W., & Kroon, M. C. (2013). Comparison of a low transition temperature mixture (LTTM) formed by lactic acid and choline chloride with choline lactate ionic liquid and the choline chloride salt: Physical properties and vapour–liquid equilibria of mixtures containing water and ethanol. *RSC Advances, 3*(45), 23553–23561.
9. Anitha, M., Kamarudin, S. K., & Kofli, N. T. (2016). The potential of glycerol as a value-added commodity. *Chemical Engineering Journal, 295*, 119–130.
10. Hou, X., Wang, Z., Sun, J., Li, M., Wang, S., Chen, K., et al. (2019). A microwave-assisted aqueous ionic liquid pretreatment to enhance enzymatic hydrolysis of Eucalyptus and its mechanism. *Bioresource Technology, 272*, 99–104.
11. Jiang, Z., Liu, Z., Fei, B., Cai, Z., Yu, Y., & Xe, L. (2012). The pyrolysis characteristics of moso bamboo. *Journal of Analytical & Applied Pyrolysis, 94*, 48–52.
12. Li, M., L-j, W., Li, D., Cheng, Y.-L., & Adhikari, B. (2014). Preparation and characterization of cellulose nanofibers from de-pectinated sugar beet pulp. *Carbohydrate Polymers, 102*, 136–143.
13. Perez-Pimienta, J. A., Lopez-Ortega, M. G., Chavez-Carvayar, J. A., Varanasi, P., Stavila, V., Cheng, G., et al. (2015). Characterization of agave bagasse as a function of ionic liquid pretreatment. *Biomass & Bioenergy, 75*, 180–188.
14. Li, P., Zhang, Q., Zhang, X., Zhanga, X., Pan, X., & Xu, F. (2019). Subcellular dissolution of xylan and lignin for enhancing enzymatic hydrolysis of microwave assisted deep eutectic solvent pretreated Pinus bungeana Zucc. *Bioresource Technology, 288*, 121475.
15. Phromphithak, S., Tippayawong, N., Onsree, T., & Lauterbach, J. (2022). Pretreatment of corncob with green deep eutectic solvent to enhance cellulose accessibility for energy and fuel applications. *Energy Reports, 8*, 579–581.
16. Rigual, V., Domínguez, J. C., Santos, T. M., Rivas, S., Alonso, M. V., Oliet, M., et al. (2019). Autohydrolysis and microwave ionic liquid pretreatment of Pinus radiata: Imaging visualization and analysis to understand enzymatic digestibility. *Industrial Crops and Products, 134*, 328–337.
17. Ji, Q., Yu, X., Yagoub, A. E.-G. A., Chen, L., & Zhou, C. (2020). Efficient removal of lignin from vegetable wastes by ultrasonic and microwave-assisted treatment with ternary deep eutectic solvent. *Industrial Crops and Products, 149*, 112357.
18. Aguilar-Reynosa, A., Romaní, A., Rodríguez-Jasso, R. M., Aguilar, C. N., Garrote, G., & Ruiz, H. A. (2017). Microwave heating processing as alternative of pretreatment in second-generation biorefinery: An overview. *Energy Conversion & Management, 136*, 50–65.

Study on Enhancement of the Interfacial Water Absorption Process by an Electrostatic Field

Fangsu Fan and Yonggao Yin

1 Introduction

The worsening energy shortage and wastefulness are leading to a progressively harsher climate environment on the planet, which will significantly impact the survival and social development of all humankind. The worldwide trend towards low-carbon energy and industrial development has gained momentum, with many countries setting carbon-neutral targets [1]. Concerning the indoor environments of buildings, there exists a fundamental contradiction between the increasing demand for thermal comfort and the necessity for low-carbon technologies. As the demand for thermal comfort within buildings continues to rise, a substantial amount of energy is consumed by Heating, Ventilation, and Air Conditioning (HVAC) systems, including the regulation of temperature and humidity in public buildings, industrial facilities, and residences. The Liquid Desiccant Dehumidification System (LDDS) can be employed independently or in conjunction with a refrigeration unit to provide multifunctional humidity control, thereby enhancing energy efficiency. This approach holds vast potential and circumvents the high energy consumption associated with traditional dehumidification using compression refrigeration, as well as the environmental pollution stemming from refrigerant leakage [2]. Moreover, the regeneration of this system can be accomplished using renewable energy sources (such as solar energy) and industrial waste heat. However, the widespread implementation of LDDS faces several challenges, including limited mass transfer

F. Fan (✉)
School of Energy and Environment, Southeast University, Nanjing, China

Y. Yin
School of Energy and Environment, Southeast University, Nanjing, China

Engineering Research Center of Building Equipment, Energy, and Environment, Ministry of Education, Nanjing, China

X. Wang (ed.), *Future Directions in Energy Engineering*, Green Energy and Technology, https://doi.org/10.1007/978-3-031-62042-3_12

efficiency within dehumidifiers, significant equipment footprint, and low energy conversion efficiency, all of which impede the further development of this technology [3].

An electric field, as an external force, enhances mass transfer processes. Compared to magnetic, microwave, laser, and other high-energy fields, E-fields present multiple advantages: low energy consumption, absence of current, eco-friendliness, and widespread accessibility. In 1976, ASAKAWA et al. pioneered the exploration of E-fields' influence on heat and mass transfer. Their work unveiled E-fields' facilitative impact on heating processes and inhibitive effect on cooling processes [4]. However, the mechanism behind the enhanced heat transfer is not yet clear. Un-raveling the intricate mass transfer mechanisms between saline solutions and humid air, as well as developing novel methods, depends on in-depth studies of interphase transport mechanisms. Okuno et al. elucidated the effect of E-fields on the evaporation process of pure water using MD simulations [5]. Their study found that E-fields in the perpendicular direction significantly enhanced water evaporation, while those in the parallel direction exhibited inhibition. This phenomenon was attributed to the influence of E-field on changing the network arrangement of water molecules. An in-depth study [6], using in-situ Raman characterization and molecular dynamics simulations, found that the applied E-field significantly reduces the free energy barrier for water evaporation, thereby markedly increasing the flux of evaporating water molecules. These findings bolster our confidence in applying E-fields to enhance mass transfer processes at the solution-air interface.

Regarding the transfer of water molecules at the interface of highly concentrated salts, some scholars have conducted relevant studies and discovered that the structure and transport properties of the liquid-vapor interface are critical for understanding the mechanism of molecular transport resistance at the liquid-vapor interface [7, 8]. Researchers have observed variations in the distribution of interfacial particles at the liquid-vapor interface in aqueous solutions of sodium salts containing different anions, with ions showing enrichment or depletion in the near-surface region [9]. MD simulation studies have revealed that the electro-static attraction of halogenated ions is most significant during the initial phase of absorption, with the release process being most hindered. Once the system reaches equilibrium, the attraction of metal ions has the greatest impact. These findings have prompted the exploration of using external E-fields to adjust the orientation of water molecules and ion distribution in the solution, thereby facilitating the transfer of water molecules between the liquid-air interface. In this work, water vapor molecules were generated from a high-temperature pure water box and continuously absorbed by the concentrated salt solutions. Lithium chloride was selected in this study, which is a hygroscopic salt commonly used for solution dehumidification. An electric field was applied in the z-axis direction of the system to explore the effect of the electric field on the absorption rate, the water molecule absorption condition at the interface and the hydrogen bonding network at the solution surface were explored. This study delves into the molecular-level facilitative role of an electric field under humid conditions, providing valuable mechanistic explanations for further enhancing moisture absorption processes using E-fields.

2 Methodology

MD simulations are widely employed to investigate mass transfer phenomena between liquid and vapor phases. To begin, the geometric model was established for simulating the absorption process and a simulation box was created containing a concentrated salt solution with water molecules, anions, and cations. The system was run under energy minimization, constant temperature, constant volume (NVT), and constant temperature, constant pressure (NPT) to achieve equilibrium. The simulation box was extended along the ±z direction to create two vapor-phase regions. Subsequently, after running NVT simulations for 5 nm, a two-phase system was obtained representing the vapor-liquid interface of the solution, designated as box_1. The dimensions of box_1 were 5 nm in the x and y directions and 16 nm in the z direction. In the same manner, a two-phase system was established for pure water, labeled box_2, with the same dimensions as box_1. The left end of box_1 was connected to the right end of box_2 to form the final simulation system, a process facilitated by Python. The concentrations under investigation fell within the typical range used for LDDS, a molar mass of 11.35 mol·kg^{-1}. Box_1 contained 2000 water molecules and 409 ion pairs, and box_2 contained 2000 water molecules. This process yielded the geometric model for the absorption process, with dimensions of 5 nm in length and width, and 32 nm in total height, as shown in Fig. 1.

The combined system underwent 500 picoseconds of NVT simulations, with coordinate and energy data output every 1 ps, and recorded in trajectory energy files. To obtain more accurate absorption data, we performed simulations with 15 sets of initial structures and calculated the averages. In these simulations, the temperature of LiCl solution was maintained at 300 K, while water in box_2 was held at 370 K to ensure constant water vapor flow, and the pressure throughout the process was held at 1 atm. An E-field was applied to each molecule to simulate the effects of the field on the molecules. The range of E-field strengths studied was from 0 to 1 V·nm^{-1}, which is two orders of magnitude higher than actual values, to compensate for the limited simulation time.

All simulations were performed using GROMACS, applying periodic boundary conditions (PBC) in all directions. All atoms or ions were treated as Lennard-Jones (LJ) particles with charges. The interactions potentials between two particles were correlated by the summation of LJ potential V_{LJ} and Coulomb potentials V_c as follows:

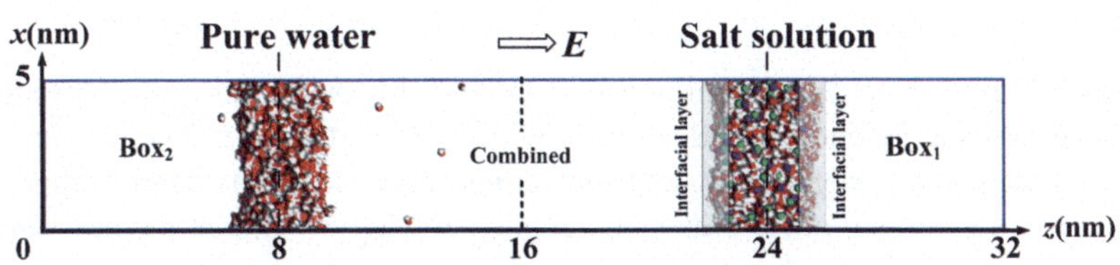

Fig. 1 Snapshot of the y-direction front view of the transport system

Table 1 Potential parameters of particles in aqueous solution

Atoms/ions	σ, Å	ε, KJ·mol^{-1}	q, e	r, Å	θ_{HOH}, °
Li^+	1.505	0.6902	1	–	–
Cl^-	4.401	0.4185	−1		
O (H_2O)	3.166	0.6502	−0.8476	O-H: 1	109.47
H (H_2O)	0.000	0.0000	0.4238		

$$V_{LJ}(r_{ij}) = 4\varepsilon_{ij}\left[\left(\frac{\sigma_{ij}}{r_{ij}}\right)^{12} - \left(\frac{\sigma_{ij}}{r_{ij}}\right)^{6}\right] \tag{1}$$

$$V_c(r_{ij}) = f\frac{q_i q_j}{r_{ij}} \tag{2}$$

where the subscripts i and j denote two different particles. ε and σ are potential parameters that vary depending on the type of atom, q represents the charges on particles, r represents the distance between particles. f is the electric conversion factor with the value of 138.9354578 kJ·mol^{-1}·nm·e^{-2}.

The simulation employed LJ potential parameters for the particles, as specified by Koneshan et al. [10] (see Table 1), a recognized force field for accurately modeling the interfacial structure and transport of concentrated lithium chloride. Constraints on water molecule bonds and angles were applied using the SETTLE algorithm, and the LJ potential constants between different particles were determined according to the Lorentz-Berthelot rule. Lennard-Jones potentials and short-range electrostatic interactions were cutoff at 12 Å, with a smoothing range between 10 and 12 Å utilizing a force-based switching function [11]. Long-range electrostatic interactions were handled with the particle-mesh-Ewald (PME) method algorithm. Particle motion integration used the leap-frog method with a 1 fs time step. The Nosé-Hoover thermostat maintained the system at the target temperature during pre-equilibration periods [12]. The simulated values for solution density and surface tension were 1121.32 kg·m^{-3} and 82.51 mN·m^{-1}, respectively. The associated errors, in comparison to experimental results of 1196.21 kg·m^{-3} and 90.31 mN·m^{-1}, were 6.26% and 8.64%, respectively. It is important to note that a long-range correction, as proposed by Janecek et al. [13], was applied to the surface tension calculation.

3 Results and Discussion

This study aimed to predict the influence of the E-fields on the absorption process of water molecules in a concentrated solution within the system's intervening junction. Figure 2 depicts the system's transient state at various time points. High-temperature water continuously supplied water molecules. Owing to the attraction between anions and cations, a steady stream of water molecules approaches the interfacial layer, where they are absorbed, or they interact with and displace other water

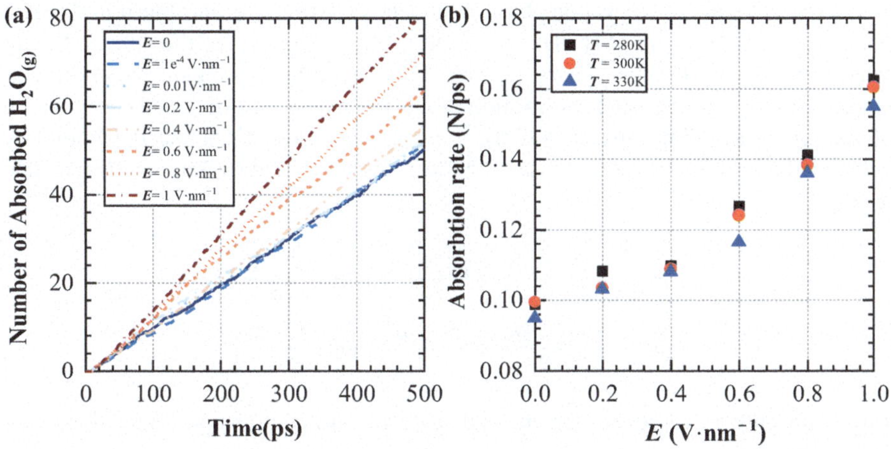

Fig. 2 Variation of (**a**) absorption amount and (**b**) the absorption rate of water molecules at the solution interface with E-field strength and temperature

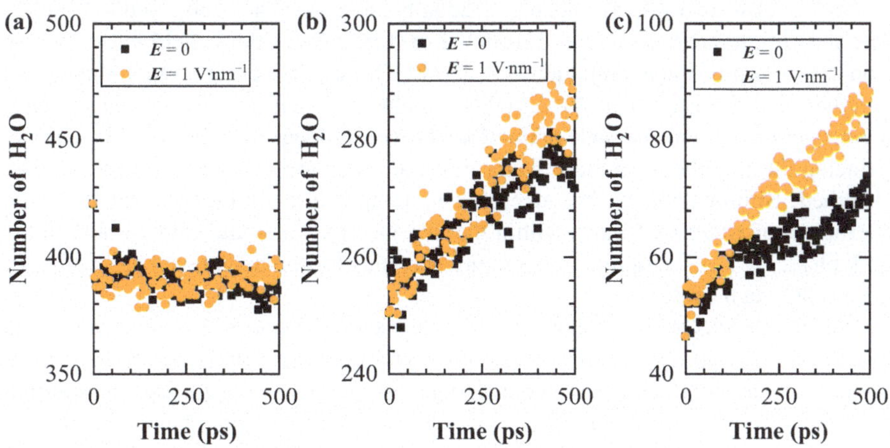

Fig. 3 Number of water molecules in (**a**) $area_1$, (**b**) $area_2$, and (**c**) $area_3$

molecules at the interface before eventually departing. We quantified the number of water molecules entering the solution from the water source in each frame, providing data on water molecule absorption quantities and flux over a 500 ps duration, as depicted in Fig. 3. Water molecules from box_2 that penetrate the interfacial layer of solution box_1 are considered absorbed water molecules.

To precisely track the ingress of water molecules into the solution, the solution box_1 was partitioned into two segments: the bulk phase, and two interfacial sides of the bulk phase. The box_1 was subdivided into slices, each with a thickness of 0.05 nm along the z-axis and then computed the average density within each of these slices to derive the local mass density distribution along the z-axis. The interfacial layers are defined by the vapor-phase boundary at 10% ρ_{bulk} and the liquid-phase boundary at 90% ρ_{bulk} [14]. To characterize the thickness of the left and right interfacial layers (the grey aera in Fig. 1), a hyperbolic tangent function was employed to fit the density distribution along the z-axis and obtained the position of the Gibbs dividing surface (GDS), $z = z_0$, were used to define the boundary between the liquid and vapor phases. The left side was fitted by Eq. 3 and right side by Eq. 4.

$$\rho(z) = \frac{\rho_L + \rho_V}{2} - \frac{\rho_L - \rho_V}{2} \tanh\left(\frac{z - z_0}{\delta}\right) \tag{3}$$

$$\rho(z) = \frac{\rho_L + \rho_V}{2} + \frac{\rho_L - \rho_V}{2} \tanh\left(\frac{z - z_0}{\delta}\right) \tag{4}$$

where, ρ_L and ρ_V represent the mass density of bulk liquid phase and the vapor phase, respectively.

Given that the absorption process continued for 500 ps and did not reach saturation, we considered the early stages of absorption as a linear process, where the absorption flux is represented by the slope of a straight line. It is evident that the quantity of absorbed water molecules gradually increases when subjected to E-fields. The absorption rate has increased from 0.1 to 0.16 N/ps as the E-field intensity rose from 0 to 1 V·nm^{-1} at a temperature of 280 K. Through temperature modulation, we ascertain that the impact of the E-fields remains pronounced. Decreasing the temperature enhances the H-bonds of interfacial water molecules on the solution surface, reducing the likelihood of these water molecules departing from the interface once absorbed. Simultaneously, the decrease in temperature diminishes the diffusion coefficient of water molecules within the interfacial layer of the solution and in the bulk phase. This reduction in temperature accounts for the heightened absorption of interfacial water molecules.

Water networks at solution interfaces are intricately connected to the creation and dissolution of hydrogen bonds (H-bonds). The quantities of H-bonds in various regions of the interfacial area were derived from Eq. 5. We divide the solution interface into the inner interfacial layer close to the liquid phase, the middle interfacial layer, and the outer interfacial layer close to the vapor phase, denoted as area$_1$, area$_2$ and area$_3$, respectively, and the results shown below are averaged over the two interfaces.

$$(2N_{00} + N_{01})/N_{\text{water}} = N_{\text{Hbond}} \tag{5}$$

Where, N_{00} is H-bonds between water molecules in the layer and N_{01} is H-bonds formed between water molecules in the layer and other water molecules.

As depicted in Fig. 3, the quantity of water molecules in area$_1$ remains relatively stable throughout the absorption process. The presence of the E-fields has minimal impact on the number of water molecules in this area, primarily due to their similar properties to those in the liquid-phase bulk. The stronger intermolecular interactions effectively counteract the influence of the E-fields. Conversely, the number of water

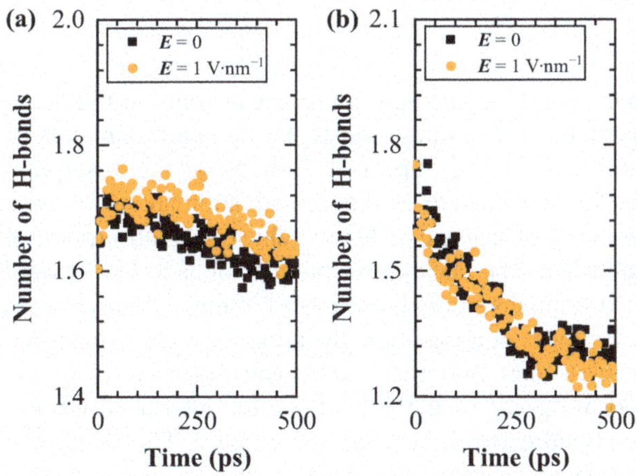

Fig. 4 Number of H-bonds in (**a**) $area_1$, (**b**) $area_2$, and (**c**) $area_3$

molecules in $area_2$ and $area_3$ exhibits a significant increase over time as absorption proceeds. This indicates that a substantial number of water molecules initially aggregate in these regions at the outset of the absorption process. The impact of the E-fields on the rate of water molecule accumulation in these two regions was clearly evident. Concerning the number of H-bonds (see Fig. 4), the E-fields induced an increased diffusion of water molecules from the liquid phase to the vapor phase side, resulting in a reduction in the total number of H-bonds within the interfacial region. The disruption of the H-bond networks by the E-fields made it easier for water molecules to enter or exit the solution interface.

The gradual pace of the absorption process can be attributed to the strong H-bonds network of water molecules at the interface, which acts as a shield, preventing the ions beneath the interface from rapid interaction. With the introduction of the E-fields, there is a noticeable weakening of the mutual attraction between water molecules within the interfacial layer. This weakening is evident from the changes in H-bonds. While the middle and outer layers of the interfacial region display an increased presence of water molecules influenced by the E-fields, a significant portion of these additional water molecules fails to establish extensive H-bonds. This observation highlights the impact of the E-fields. Consequently, it can be inferred that the primary effect of the E-fields is the disruption of the H-bonds network at the interface. Additionally, the E-fields decrease the surface tension of the interface, shown in our previous work [15]. Surface tensions were calculated from the results of the long-time NVT relaxation of box_1, yielding surface tensions of 77.85 (for $E = 0$) and 57.16 (for $E = 1$ V·nm^{-1}), respectively.

4 Conclusion

The utilization of E-fields to expedite the absorption process at solution interfaces represents an innovative approach. In this study, molecular dynamics simulations were employed to predict and analyze the impact of E-fields on the absorption process and its outcomes. The key conclusions of this investigation are as follows:

Firstly, the theoretical feasibility of using E-fields to enhance the absorption of lithium chloride has been established. The rate of absorption exhibits a clear upward trend within the range of E-field strengths studied, from 0 to 1 $V{\cdot}nm^{-1}$. Furthermore, it was observed that in the early stages of absorption, the presence of an E-fields led to a higher intake of water molecules into the middle and outer layers of the interfacial region. The primary influence of the E-fields on the interfacial layer is the disruption of the hydrogen bond network near the vapor phase. This disruption facilitates the entry of water molecules into the system.

References

1. Shang, W. L., & Lv, Z. (2023). Low carbon technology for carbon neutrality in sustainable cities: A survey. *Sustainable Cities and Society., 92*, 104489.
2. Yin, Y., Qian, J., & Zhang, X. (2014). Recent advancements in liquid desiccant dehumidification technology. *Renewable and Sustainable Energy Reviews., 31*, 38–52.
3. Su, W., Lu, Z., She, X., Zhou, J., Wang, F., Sun, B., & Zhang, X. (2022). Liquid desiccant regeneration for advanced air conditioning: A comprehensive review on desiccant materials, regenerators, systems and improvement technologies. *Applied Energy, 308*, 118394.
4. Asakawa, Y. (1976). Promotion and retardation of heat transfer by electric fields. *Nature, 261*, 220–221.
5. Okuno, Y., Minagawa, M., Matsumoto, H., & Tanioka, A. (2009). Simulation study on the influence of an electric field on water evaporation. *Journal of Molecular Structure: THEOCHEM., 904*(1–3), 83–90.
6. Fei, J., Ding, B., Koh, S. W., Ge, J., Wang, X., Lee, L., Sun, Z., Yao, M., Chen, Y., Gao, H., & Li, H. (2021). Mechanistic Investigation of Electrostatic Field-Enhanced Water Evaporation. *Advanced Science, 8*(18), 2100875.
7. Jungwirth, P., & Tobias, D. J. (2002). Ions at the air/water interface. *The Journal of Physical Chemistry. B, 106*(25), 6361–6373.
8. Björneholm, O., Hansen, M. H., Hodgson, A., Liu, L.-M., Limmer, D. T., Michaelides, A., Pedevilla, P., Rossmeisl, J., Shen, H., Tocci, G., Tyrode, E., Walz, M.-M., Werner, J., & Bluhm, H. (2016). Water at interfaces. *Chemical Reviews, 116*(13), 7698–7726.
9. Chen, T., Yin, Y., Zhang, Y., & Zhang, X. (2020). A molecular dynamics study on interaction contributions of components in liquid-vapor systems between LiBr aqueous solutions and air during absorption. *Applied Thermal Engineering., 166*, 114732.
10. Koneshan, S., Rasaiah, J. C., Lynden-Bell, R. M., & Lee, S. H. (1998). Solvent structure, dynamics, and ion mobility in aqueous solutions at 25 °C. *The Journal of Physical Chemistry. B, 102*(21), 4193–4204.
11. Steinbach, P. J., & Brooks, B. R. (1994). New spherical-cutoff methods for long-range forces in macromolecular simulation. *Journal of Computational Chemistry, 15*(7), 667–683.
12. Evans, D. J., & Holian, B. L. (1985). The Nose–Hoover thermostat. *The Journal of Chemical Physics, 83*, 4069.

13. Janeček, J. (2006). Long range corrections in inhomogeneous simulations. *The Journal of Physical Chemistry. B, 110*(12), 6264–6269.
14. Taylor, R. S., Dang, L. X., & Garrett, B. C. (1996). Molecular dynamics simulations of the liquid/vapor interface of SPC/E water. *The Journal of Physical Chemistry, 100*(28), 11720–11725.
15. Fan, F., Yin, Y., Xu, G., Lei, B., & Chen, W. (2024). Interfacial structure and transport properties of concentrated lithium chloride solutions under an electrostatic field. *Colloid Surface A, 685*, 133217.

Performance Assessment of Co-MnO_2/C Cathodic Catalyst in Pipe-Type Microbial Fuel Cells

Trang Nakamoto, Soichiro Hirose, Kentaro Inukai, and Kozo Taguchi

1 Introduction

Recently, there has been a global surge in the desire to employ eco-friendly and sustainable energy sources. Therefore, there is a rising need for power generation methods that alleviate environmental impact [1]. Consequently, microbial fuel cells (MFCs) have garnered significant interest because MFCs can generate electricity based on a biological process.

An MFC includes an anode and a cathode. The biofilm of electrogenic bacteria on the anode decomposes organic matter to grow and generate electrons and protons. The electrons are transferred to the anode and go to the cathode through an external circuit. The protons travel to the cathode surface through the water medium. At the cathode surface, the electrons and protons are reduced by oxygen molecules to form water [2]. Thanks to this operation principle, MFCs are considered a sustainable power generation method. Nevertheless, MFCs are grappling with a fundamental issue of limited power output [3].

In the pursuit of making practical MFCs, there is a need to simultaneously enhance the power output and reduce the cost of the electrodes. Typically, platinum catalysts are employed to activate the oxygen reduction reaction (ORR) at the cathode and boost MFC power production [4]. Nevertheless, the high expense and limited availability of platinum pose challenges to its application in practical MFCs. Consequently, there is a growing interest in exploring alternative catalysts to replace platinum [5].

Cathodic catalysts made from manganese dioxide (MnO_2) have been studied as a low-cost substitute for platinum [5]. MnO_2 enables high ORR performance for the cathode [6]. In addition, the research conducted in Xia et al. [7] demonstrated that the

T. Nakamoto (✉) · S. Hirose · K. Inukai · K. Taguchi
Ritsumeikan University, Kusatsu, Shiga, Japan
e-mail: n-trang@fc.ritsumei.ac.jp

X. Wang (ed.), *Future Directions in Energy Engineering*, Green Energy and Technology, https://doi.org/10.1007/978-3-031-62042-3_13

insertion of cobalt within the MnO_2 structure results in enhancements in surface area, conductivity, and catalytic performance at the air cathode of aluminum-air batteries.

In this research, we investigated how effective the Co-MnO_2/C catalyst is in improving the performance of the air cathode in a pipe-type MFC. Co-MnO_2/C synthesis does not require processes such as hydrothermal synthesis and can be made by a simple oxygen reduction reaction, thus lowering the cost of making it. To assess the impact of the catalyst on the cathode performance, an experiment using the MFCs with and without the catalyst in their air cathodes was conducted.

2 Materials and Methods

2.1 Co-MnO_2/C Catalyst Synthesis

The preparation of the catalyst was as follows [7]. A 0.1 M potassium permanganate (Wako Pure Chemical Industries, Ltd) solution was prepared in deionized water. Then 0.1 M cobalt nitrate hexahydrate (Wako Pure Chemical Industries, Ltd) was prepared in deionized water. After that, the two solutions were slowly mixed together by a magnetic stirrer at 70 °C for 1 h. Then 0.5 g of carbon black was added to the mixture and stirred further for 10 min. Finally, the resulting Co-MnO_2/C catalyst was obtained from the solution by centrifugation and dried at 60 °C for 20 h in a drying oven.

2.2 Pipe-Type MFC Fabrication

The MFC was fabricated inside a pipe. The MFC structure is shown in Fig. 1a. The anode and cathode were attached to an electrode holder, which was attached inside a PVC pipe.

The cathodic material was prepared by mixing 1 g Co-MnO_2/C catalyst in 10 ml carbon nanotube (CNT)-based ink (N7006L CNT coating liquid provided by KJ Specialty Paper Co., Ltd.). Then, it was coated on a urethane foam (5 mm thickness) attached to the cathode holder and dried at 40 °C for 24 h. In addition, the cathode without the catalyst was prepared similarly except adding the Co-MnO_2/C catalyst powder.

Similarly, an anodic coating solution was prepared by mixing 2 g carbon fiber and 2 g activated carbon powder in 10 ml CNT ink. After that, the anode was fabricated by coating the anodic solution on a urethane foam attached to the anode holder and dried at 40 °C for 24 h.

The anode and cathode holders were attached to the inside of the pipe. Two current connectors of the anode and cathode made of stainless steel wire were put outside of the pipe through two small holes. Then, the holes were sealed by a polymer bond, as shown in Fig. 1b.

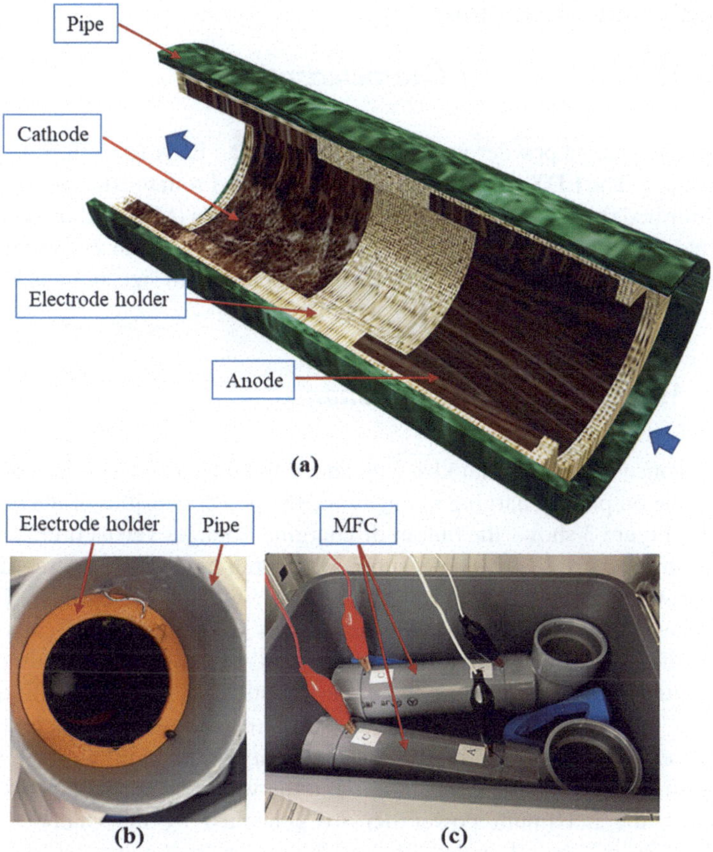

Fig. 1 Pipe-type MFC structure. A schematic of the MFC (**a**), a picture of the electrodes inside the pipe (**b**), and a picture of the experimental setup (**c**)

2.3 *Evaluation Methods*

The fabricated catalyst was characterized by an energy-dispersive X-ray spectroscopy (EDX).

The MFCs were evaluated at a temperature of 25–26 °C for more than a month in an incubator (Fig. 1c). Two types of MFC with the cathode with and without the Co-MnO$_2$/C catalyst were compared and evaluated. An external 10 kΩ resistor was connected between the two electrodes of the MFC to discharge it continuously. In addition, the power density was measured by changing the external resistor from high to low values and recording the discharging voltage at each resistor.

Luria-Bertani (LB) medium (10 g/L glucose, 10 g/L tryptone, 5 g/L yeast extract, 10 g/L NaCl) was used to simulate the wastewater. The chemical oxygen demand (COD) of the input solution was measured by a standard method using HATCH DR6000 system.

3 Results and Discussions

3.1 Co-MnO$_2$/C Catalyst Characterization

The fabricated catalyst powder was characterized by an EDX to confirm its elemental composition. The EDX spectrum is shown in Fig. 2. From the diffraction peaks, it can be confirmed that carbon (C), cobalt (Co), manganese (Mn), and oxygen (O) are the main elemental compositions of the catalyst. Moreover, the inset of Fig. 2 shows a picture of the fabricated catalyst, which is microparticle in size and black in color.

3.2 Co-MnO$_2$/C Catalyst Evaluation

The performance of the two cathodes with and without the catalyst was evaluated by comparing the output discharging voltages and the power densities of the two MFCs using them. Figure 3 shows the output discharging voltages versus time, and Fig. 4 shows the power density curves measured at the 96th hour. Both figures show that the cathode with the catalyst outperformed the one without the catalyst. The maximum discharging voltage at 10 kΩ resistor was 0.59 V with the cathode with catalyst, while that of the cathode without catalyst was only 0.05 V. Although the voltage gradually decreased, after a month, it was still over 0.4 V. In addition, when the COD of the wastewater was changed, some responses on the output voltage could be observed. This result implies the sensitivity of the biofilm on the anode to the changes in organic matter concentration in the wastewater.

Moreover, the maximum power density generated by the cathode with the catalyst was 76 mW/m^2, but that of the cathode without the catalyst was only

Fig. 2 The EDX spectrum of the Co-MnO$_2$/C catalyst. The inset is a picture of the catalyst powder

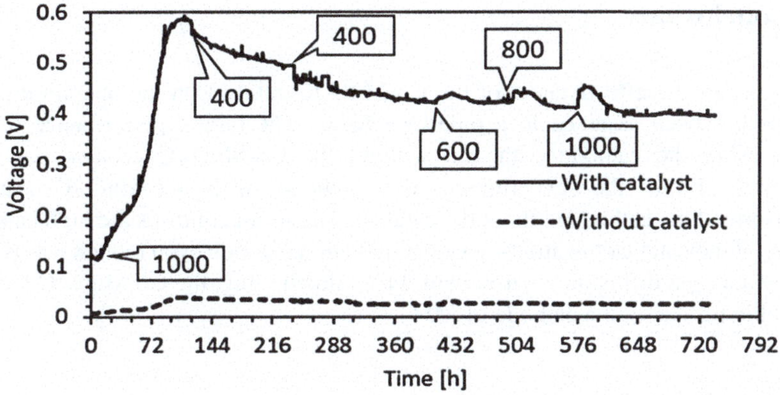

Fig. 3 Average output voltage of the MFCs using the cathodes with and without the Co-MnO_2/C catalyst. The numbers shown in the figure indicate the COD of wastewater at the time of inputting the pipes

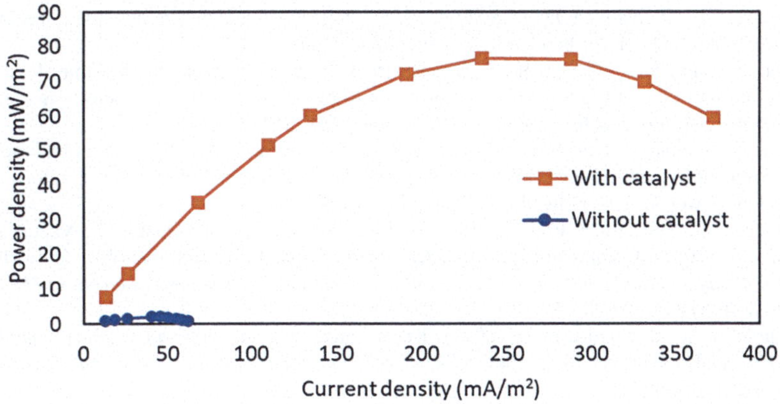

Fig. 4 The power densities of the MFCs using the cathodes with and without the Co-MnO_2/C catalyst, measured at the 96th hour

2 mW/m^2. The reason for the low performance of the cathode without the catalyst was that the cathode of the pipe-type MFC was placed inside the pipe without direct contact with open air. Consequently, the cathode without the catalyst could not perform well in such a low concentration of the diffusion oxygen in the wastewater. On the other hand, despite such low-oxygen diffusion conditions, the cathode with the catalyst performed much better. This result confirms the impact of the Co-MnO_2/C catalyst on the reduction reaction at the cathode of the MFC.

4 Conclusions

In this paper, the effectiveness of the Co-MnO_2/C catalyst on the operation of the cathode has been assessed in a pipe-type MFC. The output performance of two MFCs using the cathode with and without the Co-MnO_2/C catalyst has been compared. From the experiment, the cathode with Co-MnO_2/C catalyst outperformed the cathode without the catalyst. The result confirms the high catalytic activity of the catalyst toward the oxygen reduction reaction even though it was used in a low oxygen diffusion environment. From this finding, the Co-MnO_2/C catalyst is suitable for use in the cathode of MFC.

References

1. Tsangas, M., Papamichael, I., & Zorpas, A. (2023). Sustainable energy planning in a new situation. *Energies, 16*(4), 1626.
2. Nguyen, H. U. D., Nguyen, D. T., & Taguchi, K. (2021). A novel design portable plugged-type soil microbial fuel cell for bioelectricity generation. *Energies, 14*(3), 553–558.
3. Nawaz, A., Haq, I. U., Qaisar, K., Gunes, B., Raja, S. I., Mohyuddin, K., & Amin, H. (2022). Microbial fuel cells: Insight into simultaneous wastewater treatment and bioelectricity generation. *Process Safety and Environmental Protection, 161*, 357–373.
4. Kumar, S., Kumar, V., Kumar, R., Malyan, S. K., & Pugazhendhi, A. (2019). Microbial fuel cells as a sustainable platform technology for bioenergy, biosensing, environmental monitoring, and other low power device applications. *Fuel, 255, 115682.*
5. Rodríguez, J., Mais, L., Campana, R., Piroddi, L., Mascia, M., Gurauskis, J., Vacca, A., & Palmas, S. (2021). Comprehensive characterization of a cost-effective microbial fuel cell with Pt-free catalyst cathode and slip-casted ceramic membrane. *International Journal of Hydrogen Energy, 46*(51), 26205–26223.
6. Chaturvedi, A., & Kundu, P. P. (2021). Recent advances and perspectives in platinum-free cathode catalysts in microbial fuel cells. *Journal of Environmental Chemical Engineering, 9*(4), Article 105662.
7. Xia, Z., Zhu, Y., Zhang, W., Hu, T., Chen, T., Zhang, J., Liu, Y., Ma, H., Fang, H., & Li, L. (2020). Cobalt ion intercalated MnO_2/C as air cathode catalyst for rechargeable aluminum–air battery. *Journal of Alloys and Compounds, 824, 153950.*

MK4 Kiln in Brick Manufacturing in Peru: Reduction of Pollutants and Maximum Energy Efficiency for Sustainable Production

M. F. Huaman Diaz, T. T. Cochachi Chahua, J. J. Cordova Rosales, A. H. Contreras Perez, and J. V. Cornejo Tueros

1 Introduction

The artisanal manufacture of bricks on a global level, as Asians lead the production of bricks, generating 20% of them with modern mechanized technology, becoming an important challenge from an environmental perspective on a global level. In 2019, the World Health Organization (WHO) documented an alarming situation: nearly 99% of the global population experienced the influence of PM2.5 and PM10 particle pollution, causing approximately six million premature deaths each year [1]. The pollutants released by this brick industry, such as carbon dioxide and PM2.5 and PM10, contribute significantly and account for a quarter of global emissions [2]. On the other hand, research has corroborated that the conventional brick manufacturing process emits polluting substances into the environment, thus contributing to the increase in pollution globally [3]. This dynamic poses challenges both in terms of energy efficiency and environmental sustainability in the brick manufacturing industry. In this context, it is essential to explore strategies and approaches to improve energy efficiency and reduce the environmental impact of this vital industry [4]. The growing importance of environmental impact and intensive thermal energy consumption are aspects that have become increasingly important in the search for more sustainable industrial practices [5]. The manufacture of artisanal bricks using traditional kilns is a widespread practice in the construction industry, but it suffers a negative environmental impact due to the pollution generated when the brick is burned, the system of these kilns is energy inefficient because They use natural resources as fuel, thus generating particles [6]. It is important to note that the firing

M. F. Huaman Diaz (✉) · T. T. Cochachi Chahua · J. J. Cordova Rosales
A. H. Contreras Perez · J. V. Cornejo Tueros
Continental University, Huancayo, Peru
e-mail: 72744770@continental.edu.pe

X. Wang (ed.), *Future Directions in Energy Engineering*, Green Energy and Technology, https://doi.org/10.1007/978-3-031-62042-3_14

temperature varies significantly depending on the type of bricks being made. For example, ordinary bricks require a firing temperature in the range of 900 to 1050 °C, while basic bricks require a firing temperature ranging from 1400 to 1800 °C [7].

In Nepal, a continuous deterioration in air quality mainly attributable to the growth of informal artisanal brick factories that pollute and cannot improve their energy efficiency. This situation has raised significant concerns in terms of public health and quality of life for residents of the region [8]. Currently, artisanal brick factories are responsible for production in large quantities, due to the coal used to generate combustion that emits pollutants that harm the health of artisanal brick producers. Unfortunately, this activity has a significant negative impact on air quality [9]. Furthermore, during their manufacturing process, these bricks accumulate a considerable amount of energy, as it involves exposing blocks of earth to exceptionally high temperatures, which can reach up to 1100 degrees Celsius, for a period ranging between 10 and 40 h [10].

In Peru, artisanal brick factories are characterized by their old infrastructure and their inefficiency in the energy system for firing brick. This problem focuses on the emissions generated by traditional ovens that use firewood, eucalyptus branches and solid waste as a fuel source. This phenomenon not only negatively affects air quality and human health, but also has adverse implications on the environment and the economic viability of these artisanal brick factories.

2 Materials and Methods

To begin the experimental process, data was collected and the construction of the MK4 Oven was prepared to prepare the kiln for firing artisanal bricks. The main objective is to mitigate contaminants, analyze the energy and production efficiency of the bricks, from this analysis, we sought to generate an economic alternative with advantages that reduce the environmental impact, energy consumption and human health.

2.1 MK4 Oven Design Features

As shown in Fig. 1, the design of the MK4 oven presents a structure made up of four ovens, with a height of 60 cm and a width of 40 cm. The walls of each oven have dimensions of 40 cm high and 4 cm wide, contributing significantly to thermal insulation and heat resistance, ensuring efficient performance. As shown in Fig. 2, to facilitate heat transfer, lower ducts with a width of 8 cm are incorporated. These ducts play an important role in evenly distributing heat inside the ovens and ensuring ideal conditions. The chimney has a diameter of 30 cm and a height of 40 cm, it plays a key role in the system. Connected to the ovens through upper ducts, it facilitates air flow and helps start the cooking process.

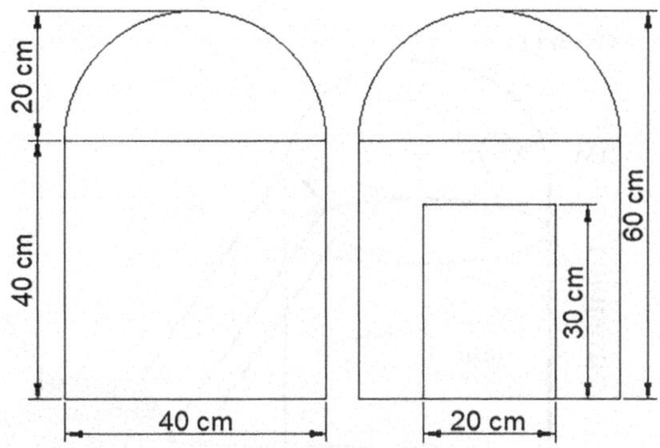

Fig. 1 Front view of the MK4 oven. (Source: self made)

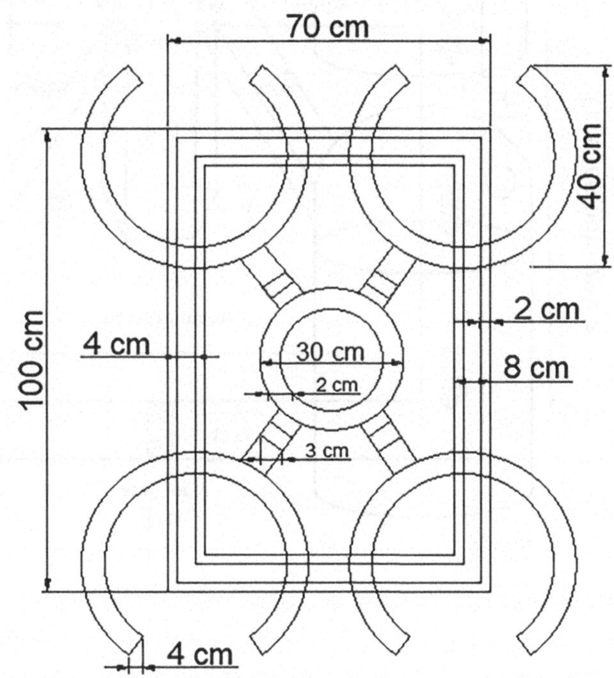

Fig. 2 Plan view of the MK4 oven. (Source: self made)

2.2 *Brick Firing Process in the MK4 Kiln*

As shown in Fig. 3, the brick firing process in the ovens (1) and (2) begins with two industrial copper torches (15) and (16), of 2050 BTU/h (approximately 0.60 kW), fed with propane gas. Which are connected to the holes designed in the upper part of the oven, so that the nozzle of the torch can enter. These torches trigger the combustion essential for firing the brick. To do this, it is necessary to have a 220 V axial fan, located in the chimney (14), with a maximum wind speed of 6.4 m/s, generating air flow in the system. This fan directs the air through the

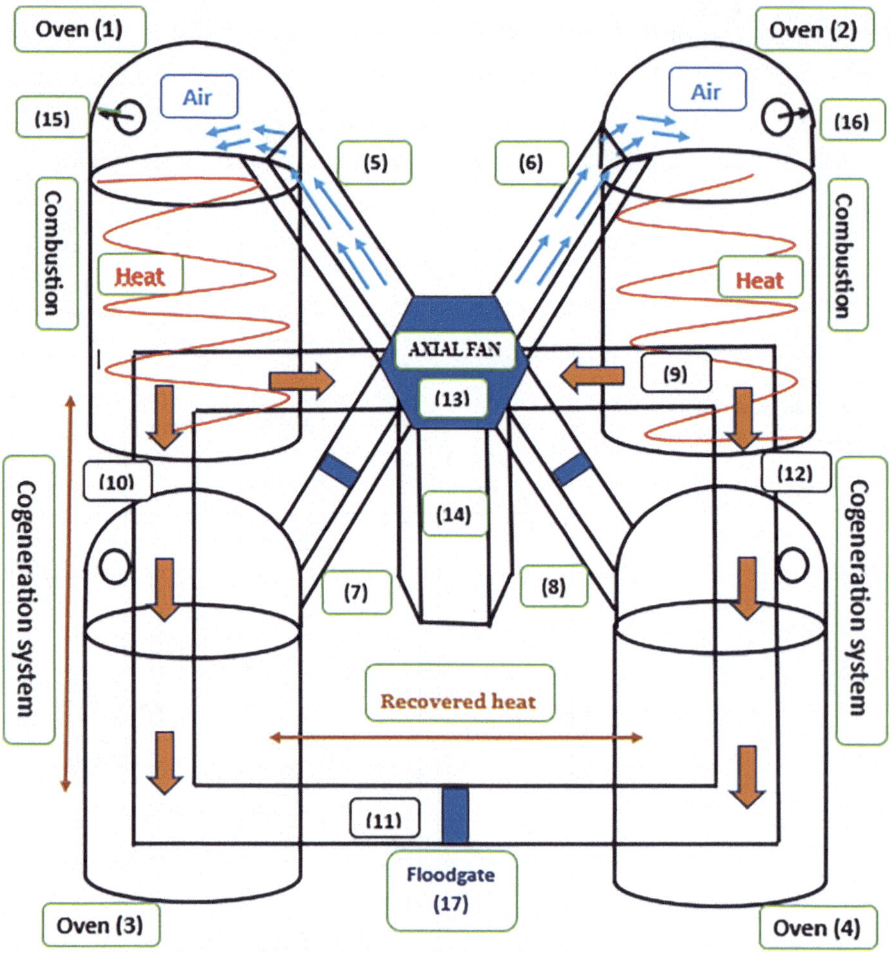

Fig. 3 MK4 furnace cogeneration system. (Source: self made)

upper ducts (5), (6), (7) and (8), according to the ovens that will begin the brick firing process. The air supply to the ovens (1) and (2) is carried out through the upper ducts (5) and (6), to start combustion in the ovens, starting the brick firing process.

2.3 *Sedimentation of Particulate Matter Through Gravitational Forces*

As shown in Fig. 3, these negative impacts affect throughout the manufacturing process. To address this, a channeling system was implemented that directs contaminants through lower conduits (9), (10), (11) and (12), which adhere particles to the raw bricks and conduits. Through the gravitational forces that sediment the particles. This system seeks to significantly reduce the emission of pollutants, mitigating the environmental impacts associated with brick manufacturing. The strategy focuses on sedimenting pollutants by gravitational force before their release, effectively reducing particles.

2.4 Energy Efficiency Through the Cogeneration System

As shown in Fig. 3, during the cooking phase, the heat generated by the ovens (1) and (2) is channeled through the lower duct (9), initiating a convection heat exchange system that generates a positive impact on both the cooking time. as in the uniformity of temperature during the brick manufacturing process. This system not only optimizes thermal performance, but also contributes to a more homogeneous temperature distribution, which is essential to obtain high-quality final products. Energy efficiency is a key priority in this process, since all the heat generated by the furnaces (1) and (2) is channeled through the lower ducts (10) and (12) as part of a cogeneration system. This design guarantees a more effective use of thermal energy, improving sustainability and reducing environmental impact. Meanwhile, kilns (3) and (4) operate in parallel, loaded with raw bricks. Once the cooking cycle in ovens (1) and (2) has been completed, the system restarts with ovens (3) and (4).

3 Result

3.1 Firing Brick in the MK4 Kiln

As shown in Fig. 4, temperature measurements during firing in ovens (1) and (2) reveal important data. In the first 20 min, the temperature reached 500.6 °C, marking the beginning of the thermal process. Even more impressive is the acceleration, reaching a maximum of 698.4 °C in 80 min, demonstrating exponential heat growth. This implies an efficient heating and thermal insulation system, which contributes to heat conservation. At the end of the 160 min, the temperature registered 995.6 °C, highlighting the effectiveness of thermal insulation and the importance of energy efficiency in the kitchen. Finally, the importance of heat exchange by convection to achieve a more homogeneous distribution of heat inside the oven.

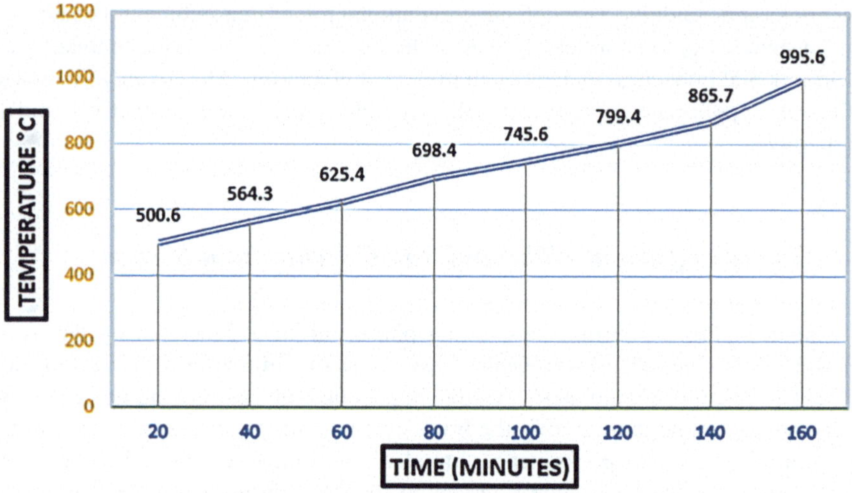

Fig. 4 Result of oven temperature and cooking time. (Source: self made)

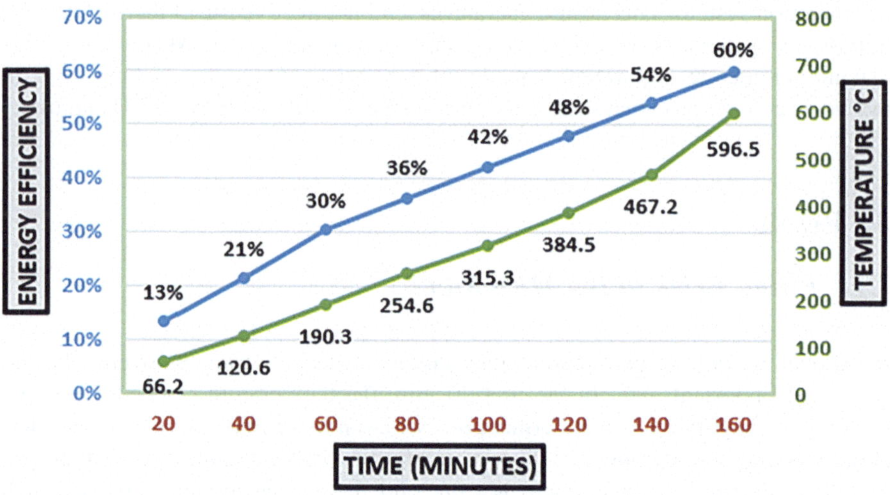

Fig. 5 Result of energy efficiency in the MK4 oven. (Source: self made)

3.2 Increase in Energy Efficiency Through the Cogeneration System

As shown in Fig. 5, the brick firing process in kilns (1) and (2) of the MK4 kiln, an innovative cogeneration system is used to maximize energy efficiency. At 20 min, the second oven shows a notable temperature of 66.2 °C, achieving an efficiency of 13% and demonstrating the effectiveness of the system. After 80 min, the temperature reaches 254.6 °C, with an energy efficiency of 36%, highlighting the efficiency of the second oven compared to the total heat generation. Upon completion of the 160 min of cooking, a temperature of 596.5 °C is achieved, with an energy efficiency of 60%, promoting a continuous process between the four ovens.

3.3 Particulate Matter During Brick Firing

Fountain: Supreme Decree No. 003-2017-MINAM—PERU—Environmental Quality Standards (ECA) for Air Are Approved [11]

As shown in Fig. 6, the analysis of the emissions generated during the firing of the brick in the MK4 kiln reveals results in terms of air quality. The levels of suspended particles (PM 2.5 and PM10) obtained, with 23 μg/m^3 and 60 μg/m^3 respectively, are below the limits established in Table 1, environmental regulations which establish a maximum of 50 μg/m^3 for PM2. 5 and 100 μg/m^{3}for PM10. This compliance indicates that the MK4 furnace system is efficient and meets environmental quality standards. It is important to highlight that these results were obtained through specific monitoring designed to measure various atmospheric pollutants, using portable and versatile equipment such as the AEROQUAL SERIE 500.

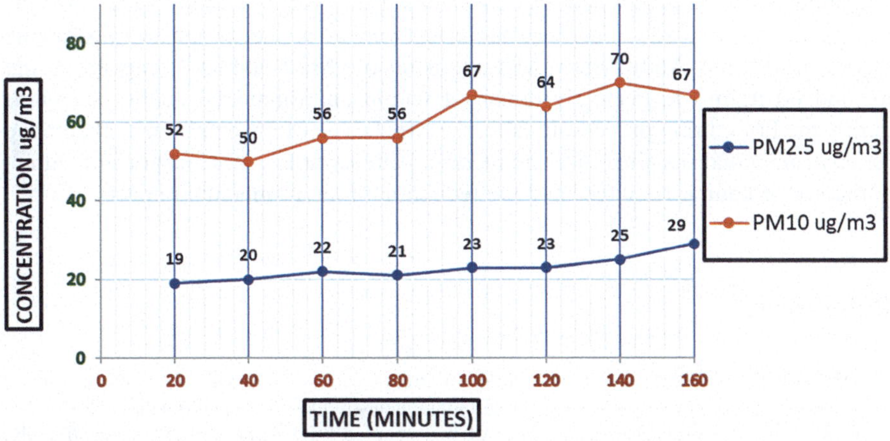

Fig. 6 Results of PM 2.5 and PM10 emissions from the MK4 oven. (Source: self made)

Table 1 Environmental quality standards for air

Parameters	Period	Worth [mg/m^3]
Material particles with diameter less than 2.5 microns (PM2.5)	24 h	50
	Annual	25
Material particles with diameter less than 10 microns (PM10)	24 h	100
	Annual	50

4 Discussion

Artisanal brickyards, in contrast to modern facilities, lack advanced control systems to mitigate polluting emissions and have significantly lower levels of efficiency in brick production. As a result, a source of environmental concern of considerable magnitude arises that significantly affects the ecosystem. The low energy efficiency in artisanal brickyards translates into an inefficient use of resources, such as the wood used to bake the bricks. Burning wood in primitive kilns entails high fuel consumption and consequently emits large amounts of greenhouse gases, contributing to climate change and air pollution [12].

5 Conclusion

In conclusion, the MK4 oven is an innovative system that consists of heat exchange by convection, this is because the air inside the ovens becomes hotter, and therefore less dense, in the cooking process. of handmade bricks. Ovens (1) and (2) reached 995.6 °C in 160 min, showing faster and more uniform cooking, essential for the quality of the product. An energy efficiency of 60% was also obtained through the cogeneration system in 160 min. These results reduce fuel consumption, reducing operating costs and greenhouse gas emissions. Likewise, heat recovery at 596.5 °C in furnaces (3) and (4) underlines the effectiveness of the design in maximizing thermal utilization. Additionally, Measurements of PM2.5 and PM10 register 23 $\mu g/m^3$ and 60 $\mu g/m^3$, respectively remained within environmental quality standards during the brick firing process. Finally, the MK4 oven offers an efficient alternative for environmental sustainability and marks a path towards a more responsible future in the use of natural resources and the reduction of environmental impacts.

References

1. OMS. (2022). *Ambient (outdoor) air pollution*. Press Center. Online 2022. Available at: https://www.who.int/es/news-room/fact-sheets/detail/ambient-(outdoor)-air-quality-and-health
2. Chávez, R. H. (2021). Energy analysis of two coupled brick kilns: An experimental study. *Clean Technologies and Environmental Policy. Online., 23*(8), 2235–2249. https://doi.org/10.1007/s10098-021-02130-y
3. Alabduljabbar, H., Benjeddou, O., Soussi, C., Khadimallah, M. A., & Alyousef, R. (2021). Effects of incorporating wood sawdust on the firing program and the physical and mechanical properties of fired clay bricks. *Construction Engineering Magazine. Online., 35*(2352–7102), 1–11. https://doi.org/10.1016/j.jobe.2020.102106
4. Chavez, R. H. (2021). Energy analysis of two coupled brick kilns: An experimental study. *Clean Technologies and Environmental Policy. On-line., 23*(8), 2235–2249. https://doi.org/10.1007/S10098-021-02130-Y

5. Refaey, H. A., Alharthi, M. A., Salem, M. R., Abdel-Aziz, A. A., Abdelrahman, H. E., & Karali, M. A. (2021). Numerical investigations of convective heat transfer for lattice settings in brick tunnel kiln: Cfd simulation with experimental validation. *Thermal Science and Engineering Progress, 24*(April), 100934. https://doi.org/10.1016/J.Tsep.2021.100934
6. Zavala, M., Molina, L. T., Maiz, P., Monsivais, I., Chow, J. C., Watson, J. G., Munguia, J. L., Cardenas, B., Fortner, E. C., Herndon, S. C., Roscioli, J. R., Kolb, C. E., & Knighton, W. B. (2018). Black carbon, organic carbon and co-pollutant emissions and energy efficiency from artisanal brick production in Mexico. *Chemistry and Atmospheric Physics, 18*(8), 6023–6037. https://doi.org/10.5194/acp-18-6023-2018
7. Almesri, I. F., Alrahmani, M. A., Almutairi, J. H., & Abou-Ziyan, H. Z. (2023). Correlations of heat transfer and fluid flow data for lattice brick configurations in tunnel kilns. *Energies, 16*(9), 3631. https://doi.org/10.3390/En16093631
8. Thygerson, S. M., Beard, J. D., House, M. J., Smith, R. L., Burbidge, H. C., Andrus, K. N., Weber, F. X., Chartier, R., & Johnston, J. D. (2019). Air-quality assessment of on-site brick-kiln worker housing in Bhaktapur, Nepal: Chemical speciation of indoor and outdoor PM2.5 pollution. *International Journal of Environmental Research and Public Health, 16*(21), 4114. https://doi.org/10.3390/ijerph16214114
9. Nepal, S., Mahapatra, P. S., Adhikari, S., Shrestha, S., Sharma, P., Shrestha, K. L., Pradhan, B. B., & Puppala, S. P. (2019). A comparative study of chimney emissions from straight line and zigzag brick kilns in Nepal atmosphere. *Atmosphere, 10*(3), 1–19. https://doi.org/10.3390/atmos10030107
10. Bruno, A. W., Gallipoli, D., Perlot, C., & Mendes, J. (2019). Optimization of bricks production by earth hypercompaction prior to firing. *Journal of Cleaner Production., 214*, 475–482. https://doi.org/10.1016/j.jclepro.2018.12.302
11. They approve Environmental Quality Standards (ECA) for Air and establish Complementary Provisions. SINIA | National Environmental Information System [online], [no date]. [consultation: June 26, 2023]. Available at: https://sinia.minam.gob.pe/normas/aprueban-estandares-calidad-ambiental-eca-aire-establecen-disposiciones
12. Valdés, H., Vilches, J., Felmer, G., Hurtado, M., & Figueroa, J. (2020). Artisanal brick ovens: State of the art and future trends. *Sustainability (Switzerland), 12*(18), 1–19. https://doi.org/10.3390/su12187724

Analyzing Supply Chain Structures with Petri Nets: A Composition-Based Approach

Olof M. A. Saif, Abobaker M. Q. Farhan, Mueataz M. A. Saif, and Mujahed Farhan

1 Introduction

In recent times, significant advancements have been observed in areas such as goal and result management within total quality control, followed by total quality management, business process re-engineering, knowledge management, and ultimately, supply chain management. The prominence of supply chains has risen due to the potential cost savings that can be achieved through an organization's efforts, as well as the vast opportunities emerging from breaking down barriers between organizations, their customers, and suppliers. Supply chain management focuses on effectively governing the movement of information, materials, services, and finances throughout various activities to optimize operational efficiency. Furthermore, it involves the implementation of new tools and the alteration or adaptation of existing methods to enhance efficiency and effectiveness [1, 2].

When done correctly, successful supply chain management can help reduce costs, manage risk, and maintain or improve value added and profit margins for both customers and suppliers [3, 4]. This is why companies that are effective in supply chain management are often the most successful in today's business world. Through the use of new technologies, such as analytic and artificial intelligence, supply chain management has become more efficient and effective, allowing organizations to gain greater control over their supply chains and more accurately predict customer demand [5, 6].

O. M. A. Saif · M. M. A. Saif
School of Business Administration, Sichuan University, Chengdu, China

A. M. Q. Farhan (✉)
School of Information and Software Engineering, University of Electronic Science and Technology of China, Chengdu, China

M. Farhan
Taiz University, Taiz, Yemen

X. Wang (ed.), *Future Directions in Energy Engineering*, Green Energy and Technology, https://doi.org/10.1007/978-3-031-62042-3_15

2 Related Work

Reference	Methodology	Key findings
[7]	Developed a Petri net model of a Brazilian food supply chain case study. Performed structural and behavioral analysis	Identified bottlenecks and proposed improvements to increase efficiency and reduce waste
[8]	Proposed a generic Petri net model for supply chain risk management. Enabled simulation of different risk scenarios	Allowed evaluation of different mitigation strategies and their impact on system performance under disruptions
[9]	Applied fuzzy colored Petri nets to model uncertainties in a pharmaceutical supply chain. Conducted qualitative and quantitative analysis	Evaluated system performance under different disruption scenarios and the effects of uncertainties
[10]	Developed a hierarchical Petri net modeling approach for Industry 4.0 supply chains. Considered different information and material flow levels	Enabled end-to-end visibility, optimization, and integration of physical and information flows across supply chain entities
[11]	Proposed a Petri net-based framework to assess resilience in agri-food supply chains. Defined resilience metrics	Identified resilience strategies through simulation and sensitivity analysis under different disruption scenarios
[12]	One of the early applications of PNs for qualitative and quantitative supply chain modeling and analysis	Demonstrated suitability of Petri nets for describing complex supply chain dynamics and behaviors
[13]	A comprehensive survey of Petri net applications in supply chain management research	Established Petri nets as a popular modeling technique and identified open research areas for future work
[14]	Developed a technique to compositionally analyze modular supply chain models built with Petri nets	Allowed modular structuring, integration, and reuse of supply chain components in a composite model

3 Food Supply Chain

In the food supply chain, food moves from the stage of obtaining raw materials, then to the farmer, and then through several stages such as production, processing, distribution, retail, and finally to the consumer, as shown in Fig. 1. This process involves various stakeholders, such as producers, processors, distributors, retailers, and consumers, all of whom play an important role in ensuring the safe and efficient delivery of food to the end consumer. Moreover, the food supply chain is a complex system, with numerous variables and factors that must be taken into account in order to ensure the quality and safety of the food that is delivered.

4 Petri Nets and Compositional Structures

Petri net networks are a formalism that can be used to model serial or parallel systems. Developed by Carl Adam Petri, a German Mathematician, in the early 1960s, the Petri nets have since evolved to represent relationships between conditions and events [15]. It is now one of the largest open academic fields for development, with many complex systems, such as business processes, manufacturing systems, distributed communications and communication protocols, being modeled using Petri networks [16, 17].

The components of a Petri network include places, transitions, arcs, tokens, and markings. Places represent the system's resources and are connected to transitions, which represents. Arcs are the lines that connect places and transitions and are used to move tokens from one place to another. Tokens are the graphical representation of resources, while markings represent the number of tokens present in each place. Petri networks are a powerful tool for modeling complex systems [18, 19].

Place is often used to refer to the situation or context in which a network is distributed across multiple places, rather than being associated with only one.

Transition: can be thought of as a process of moving from one case to another.

Arc: $F \subset (P \times T) \cup (T \times P)$, is an arrow that connects two different elements in a system.

In order to build a Petri net from the food supply chain shown in Fig. 1, we developed the Petri net model. This model consists of places, transitions, and arcs, which are used to represent the various elements of the food supply chain. The places represent the different stages of the supply chain, such as production, processing,

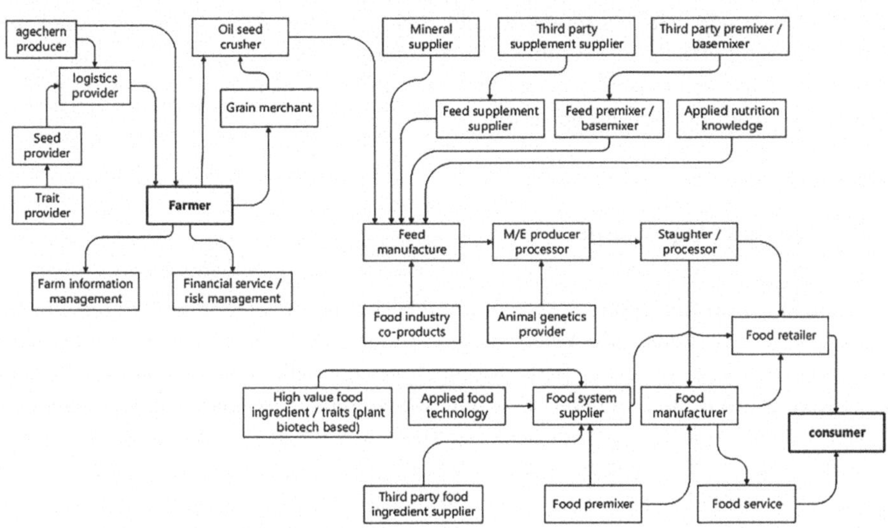

Fig. 1 Prototype food supply chain structure

Table 1 Interpretation places and transitions of Petri net model

Symbols	Description
t_n	Transportation of process from t1 to t20
p_1 to p_5	Material raw of consumers
p_6 to p_{11}	Material raw of system input
p_{12} to p_{15}	Material raw of farm
p_{16} to p_{21}	Material raw of processing
p_{22} to p_{28}	Material raw of distribution

distribution, and retail, while the transitions represent the movement of food through the supply chain. Finally, the arcs represent the relationships between the different elements of the supply chain.

Table 1 provides comprehensive descriptions for each place and transition within the Petri net model of the food supply chain. This table elucidates the various elements and interactions that contribute to the overall dynamics of the system, facilitating a better understanding of the model's structure and behavior in representing the complex food supply chain. Through this table, it is possible to gain insights into the various stages of the supply chain and the relationships between them, as well as how the different elements interact with one another to create the desired behavior.

5 Compositional Analysis Algorithm

Composition analysis is performed by the supply chain algorithm. First, each resulting incidence transitions matrix is expressed in terms of the complete algorithm obtained from the Petri net model to find each structure of the Petri net model, as shown in Fig. 2. This analysis allows for a better understanding of the structure and behavior of the system, as well as how the different elements interact with one another to create the desired behavior.

5.1 Formal Algorithm

Depending on the algorithm specification, this part defines the aimed algorithm. In the formal representation, the whole method is defined as a formal algorithm, which is presented in the following parts. Table 2 provides all symbols that used in the algorithm. Algorithm 1 obtains all result compositional structures in the Petri net model that shown in Fig. 2 by performing the following steps: First, it initializes the incidence matrix, $\#S_{sp}$, and the output result, $\#S_{out}$. Then, it iterates through all rows (places) in the incidence matrix, from P_1 to P_{28} (from the head to tail). For each row, the algorithm calculates the value of the corresponding transition, $\#t_i$. If the value of the transition is positive, the algorithm adds the corresponding row (place) to $\#S_{out}$.

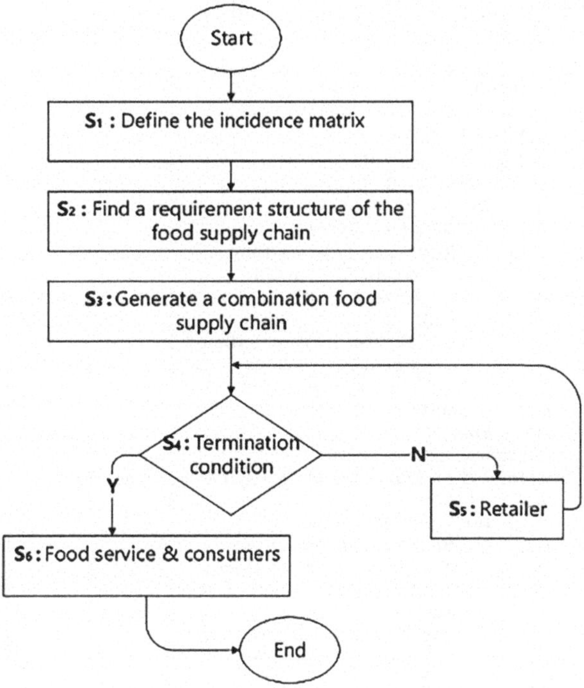

Fig. 2 Framework food supply chain

Table 2 Symbols defined in algorithm

Symbols	Explanation
X, *Y*	Initial value
#sub, #pos	Represented and stored the negative and positive values, respectively. This allows for a more efficient and accurate representation of the data, as well as a more efficient operation of the system
$\#S_{sp}$	Represented the incidence matrix
$\#t_i$, $\#P_j$	Represented the number of columns (transitions) and rows (places) of $\#S_{sp}$ and the incidence matrix using $\#t_i$ and $\#P_i$, respectively
$\#S_{out}$	Represented and stored the final result of the compositional structure of the Petri net model using $\#S_{out}$

Finally, the algorithm sets a terminate condition to ensure that the algorithm stops running once the desired outcome has been achieved, thus allowing for a more efficient and cost-effective operation of the system.

Algorithm 1 Obtain compositional Petri net model structure

```
   input : #pos; #sub; #S_sp; X = 0; Y = 0;
           SUM; #t_i; #p_j;
   output: #S_out
1  for i ∈ [0, #t_i] do
2      SUM ← 0
3      for j ∈ [0, #p_j] do
4          SUM ← SUM + i + j
5          if #S_sp(i, j) = −1 then
6              #sub(X) ← SUM
7              X ← X + 1
8          end
9          else if #S_sp(i, j) = 1 then
10             #pos(Y) ← SUM
11             Y ← Y + 1
12         end
13     end
14 end
15 for i ∈ [0, #S_sp], i ∈ [0, #pos], i ∈ [0, #sub] do
16     SUM ← 0
17     for i ∈ [0, #S_sp] do
18         SUM ← SUM + i + j
19         if #pos(i) = SUM then
20             #S_out(i, j) ← #S_sp(SUM)
21         end
22         else if #sub(i) = SUM then
23             #S_out(i, j) ← #S_sp(SUM)
24         end
25         else
26             #S_out(i, j) ← 0
27         end
28     end
29 end
```

- (Line 1–3): we created two loops to determine and separate the positive and negative values from the incidence matrix of Petri net.
- (Line 4): we created a function to calculate the summation of the row (places) values, which is used as an index for the positive and negative values.
- (Line 5–6): we checked the negative values and assigned the index of these values to the *#sub* matrix and adding one for *X* variables for iterations.
- (Line 9–10): we checked the positive values and assigned the index of these values to the *#pos* matrix and adding one for X variables for iterations.
- (Line 15–16): We performed the composition between the *#pos* and the S_{sp} matrix. In (line 15), we reset the summation values of the places in the matrix S_{sp}.
- (Line 17): we performed the composition between the *#sub* matrix and S_{sp} matrix. This composition process also helps to reduce the complexity of the system, as the composition of the two matrices eliminates the need for additional calculations.
- (Line 18): we calculated the summation of the places in the S_{sp} matrix.
- (Line 19): We checked the composition between the transition $\#t_i$) and places ($\#P_J$) from the *#pos* matrix and S_{sp} matrix, respectively.

– (Line 20): If the condition in (line 19) is true, we will assign the value of S_{sp} matrix to the S_{out} matrix, indicating that the composition is in the positive side.
– (Line 22): We checked the composition between the transition ($\#t_i$) and places ($\#P_j$) from the *#sub* matrix and S_{sp} matrix, respectively.
– (Line 25–26): If the condition in (line 22) is true, we will assign the value from S_{sp} matrix to the S_{out} matrix, indicating that the composition is in the negative side.

6 Conclusion and Future Work

This paper presents a detailed composition analysis of the supply chain, including a model specifically for the food supply chain, using the mathematical modeling language Petri nets. The composition analysis offers advantages for describing complex supply chains based on their unique requirements. The simulation and analysis results from this model can be used to refine the food supply chain, leading to more efficient and effective operations. Future work on this composition analysis could include further refining the model to more represent the operations of the food supply chain, exploring different modeling languages and methods for analyzing the supply chain, and applying the simulation and analysis results to real-world scenarios.

References

1. Choi, T. Y., & Hartley, J. L. (2019). Integrating supply chain management. *Journal of Business Logistics, 40*(1), 9–16.
2. Khan, S. A. R., Yu, Y., & Sharif, A. (2019). A review of supply chain management research in leading logistics journals: A bibliometric analysis. *Logistics, 3*(3), 19.
3. Choi, T.-S., Choi, J.-E., & Kim, J.-H. (2019). Integrating supply chain management and artificial intelligence: A review and outlook. *International Journal of Production Economics, 221*, 428–446.
4. Khan, M. K., & Khan, M. A. (2019). A review on the role of artificial intelligence in supply chain management. *International Journal of Supply Chain Management, 8*(4), 1242–1252.
5. Khan, M. S., Khan, S. A., & Khan, M. A. (2019). A review of artificial intelligence in supply chain management. *International Journal of Production Research, 57*(8), 2118–2136.
6. Rahman, A., & Rahman, M. M. (2019). A review of artificial intelligence in supply chain management. *International Journal of Production Research, 57*(9), 2297–2316.
7. Silva, J., Oliveira, M., & Souza, R.. (2022). Developing a petri net model for analyzing brazilian food supply chains. *International Journal of Supply Chain Management, 15*(3), 145–160.
8. Pohlmann, C. R., Scavarda, A. J., Alves, M. B., & Korzenowski, A. L. (2020). The role of the focal company in sustainable development goals: A brazilian food poultry supply chain case study. *Journal of Cleaner Production, 245*, 118798.
9. Jaiswal, A., Ramesh, V., Kadri, M., & Dey, N. (2021). Modeling supply chain risk management using petri nets: An application to temperature controlled supply chains. *Computers & Industrial Engineering, 157*, 107190.

10. EL Mokadem, H., Machrafi, M., & Hamadi, R. (2022). Modeling uncertainties in supply chain management using fuzzy petri nets: Application to pharmaceutical industry. *Journal of Ambient Intelligence and Humanized Computing, 13*(4), 2459–2477.
11. Liu, X., Zhang, H., & Zhang, J. (2022). Hierarchical modeling and simulation of industry 4.0-enabled supply chains based on petri nets. *IEEE Access, 10*, 1298–1310.
12. Scholten, A., & Sandin, E. (2022). Evaluating agriculture and food supply chain resilience using a petri net based framework. *Journal of Cleaner Production, 344*, 131270.
13. Karagiannis, H. D., & Giakoumakis, A. D. (2010). Modeling and analysis of supply chains using petri nets. *Digital Enterprise and Information Systems*, 55–66.
14. Shih, C.-Y., Su, Y.-S., & Jeng, L.-M. (2015). Applications of Petri nets in the modelling and analysis of supply chain systems: A literature review. *Journal of the Chinese Institute of Industrial Engineers, 32*(3), 195–208.
15. Westergaard, M., & Melcer, T. S. (2018). Modular verification of supply chain models expressed in colored petri nets. *IEEE Transactions on Industrial Informatics, 14*(2), 702–713.
16. Blackhurst, A. (2018). Supply chain risk management: A review. *International Journal of Physical Distribution & Logistics Management, 48*(3/4), 205–223.
17. Blackhurst, J., Rungtusanatham, M. J., Scheibe, K., & Ambulkar, S. (2018). Supply chain vulnerability assessment: A network based visualization and clustering analysis approach. *Journal of Purchasing and Supply Management, 24*(1), 21–30.
18. Petri, C. (1962). *Kommunikation mit Automaten* (pp. 43–72). Princeton University Press.
19. Murata, T. (1989). Petri nets: Properties, analysis and applications. *Proceedings of the IEEE, 77*(4), 541–580.

Index

X. Wang (ed.), *Future Directions in Energy Engineering*, Green Energy and Technology, https://doi.org/10.1007/978-3-031-62042-3

SPRINGER NATURE

GPSR Compliance

The European Union's (EU) General Product Safety Regulation (GPSR) is a set of rules that requires consumer products to be safe and our obligations to ensure this.

If you have any concerns about our products, you can contact us on ProductSafety@springernature.com

In case Publisher is established outside the EU, the EU authorized representative is:

Springer Nature Customer Service Center GmbH
Europaplatz 3
69115 Heidelberg, Germany

The manufacturer's authorised representative in the EU is Springer Nature Customer Service Centre GmbH, Europaplatz 3, 69115 Heidelberg, Germany. If you have any concerns regarding our products, please contact ProductSafety@springernature.com

Printed and bound by CPI Group (UK) Ltd, Croydon, CR0 4YY
30/09/2026
02225577-0006